Fractal Dimensions for Poincaré Recurrences

MONOGRAPH SERIES ON NONLINEAR SCIENCE AND COMPLEXITY

Fractal Dimensions for Poincaré Recurrences

V. AFRAIMOVICH
San Luis Potosi University, Mexico

E. UGALDE
San Luis Potosi University, Mexico

J. URÍAS
San Luis Potosi University, Mexico

ELSEVIER

AMSTERDAM • BOSTON • HEIDELBERG • LONDON • NEW YORK • OXFORD
PARIS • SAN DIEGO • SAN FRANCISCO • SINGAPORE • SYDNEY • TOKYO

Elsevier
Radarweg 29, PO Box 211, 1000 AE Amsterdam, The Netherlands
The Boulevard, Langford Lane, Kidlington, Oxford OX5 1GB, UK

First edition 2006

Library of Congress Cataloging-in-Publication Data
A catalog record for this book is available from the Library of Congress

British Library Cataloguing in Publication Data
A catalogue record for this book is available from the British Library

ISBN-13: 978-0-444-52189-7
ISBN-10: 0-444-52189-5
Series ISSN: 1574-6917

For information on all Elsevier publications
visit our website at books.elsevier.com

Printed and bound in The Netherlands

06 07 08 09 10 10 9 8 7 6 5 4 3 2 1

Preface

Research by several prominent research groups (including Penn State University, CNRS at Luminy and Ecole Polithecnique, University Simón Bolivar, our group in San Luis Potosí, etc.) has shown that the dimension theory of dynamical systems is a powerful tool to analyze the (multi)fractal behavior which appears in real systems and their mathematical models. This book is devoted to an important branch of this theory: the study of the fine (fractal) structure of Poincaré recurrences – instants of time when the system almost repeats its initial state. Because of this restriction we were able to write an entirely self-contained text including many insights and examples, as well as providing complete details of proofs. The only prerequisites are a basic knowledge of analysis and topology. Thus this book can serve as a graduate text or self-study guide for courses in applied mathematics or nonlinear dynamics (in the natural sciences).

To motivate our study of Poincare recurrences, imagine that the phase space (or invariant subset) is partitioned into different colors according to their "temperature". The orbit for initial conditions chosen in hot areas returns nearby faster than for initial conditions chosen from cold areas. More precisely, the return time for an ε ball centered on a hot initial condition is much less than for a cold initial condition. It is true even for uniformly hyperbolic systems, provided that they are "sufficiently nonlinear", i.e., their linearization depends on the point.

One can then fix a large collection of ε balls and calculate an "average" return time, and study this average as ε goes to 0. For many hyperbolic systems the average behaves as $-\gamma \log \varepsilon$ and for non-chaotic systems as $\varepsilon^{-\gamma}$, where γ is a key dimension-like characteristic obtained from the fractal dimension machinery: the dimension for Poincaré recurrences. It depends on the set of initial points we deal with, i.e., it is a function of a set. If we choose a set situated around hot spots, evidently we will obtain a dimension that is different from that around cold spots.

This has profound practical applications, for it provides a useful "measure" of how chaotic is a dynamical system within the class of chaotic dynamical systems while for nonchaotic dynamical systems it is a new and useful measure of the complexity of the orbit structure. Furthermore, given an invariant measure, it is natural to introduce a dimension of measure which equals, as usual, the dimension of the smallest set of the full measure. So, our dimension can distinguish different measures according to the behavior of Poincaré recurrences. If the measure is ergodic then the behavior of Poincaré recurrences is asymptotically the same for any ball centered at a typical point. Thus, one can obtain the dimension of measure for Poincaré recurrences, which is a global quantity, by knowing a local dimension.

We believe that the dimension could be quite useful for many applied problems. Let us emphasize that the dimension is computable and can be found numerically for specific systems. Let us mention now a problem related to synchronization phenomena. If two (or more) coupled subsystems are synchronized then their behavior in time has to be similar and since the dimension definitely reflects such a behavior, then the dimension for Poincaré recurrences in a synchronized regime has to be the same for all individual subsystems. Thus, it can serve as an indicator of synchronization.

The second problem we want to mention is the problem of fractal and multifractal features of Poincaré recurrences. In situations where a system is nonergodic and contains both, chaotic invariant subsets and subsets with zero topological entropy (such as in standard map) a normalized distribution of return times to a region behaves as follows

$$P(\tau) \sim \tau^{-\gamma}, \quad \tau \to \infty$$

(where $P(\tau)\,d\tau$ is in fact the probability to return to the region during the interval of time $(\tau, (\tau + d\tau)))$. We explained in Chapter 15 that this exponent γ is directly related to the dimension for Poincaré recurrences. So, our quantity has an important physical meaning.

The book includes figures already published in our papers. Kind permissions were received from the publishers World Scientific, Discrete and Continuous Dynamical Systems and The American Physical Society for the reproduction of the following figures of this book: Figures 3.4 and 3.5 from reference [52], Figure 15.3 from reference [12] and Figures 16.3–16.6 from reference [13].

Contents

Chapter 1

Introduction

Deterministically generated processes in nature and industry are modeled by dynamical systems. For dissipative systems mathematical images of established motions are attractors, invariant attracting sets in phase spaces. Geometry of attractors reflects some features of a motion. So, for systems with continuous time, a point in the phase space corresponds to an equilibrium state, a closed curve (a limit cycle) corresponds to a periodic motion, a torus – to a quasi-periodic motion, and generally, a strange attractor – to a chaotic oscillation.

The term "strange attractor" was introduced by mathematicians in 1970s to emphasize the fact that the attractor cannot be represented as a finite union of smooth curves or surfaces. It looks like a very bizarre, non-regular subset of the phase space. Such sets were known to be called (because of B. Mandelbrot [80]) the fractal sets. They have non-integer fractal dimension (the Hausdorff or the box dimension), and the fractionality is caused by the presence of a Cantor-like structure in some direction inside the fractal set.

Fractal sets appear in different field of science and nature, but what is the most important for us is that they serve as attractors for dynamical systems with chaotic behavior of orbits.

Usually a dynamical system is determined by a local rule, a system of ordinary differential equations or a mapping of the phase space into itself. What is surprising is that an inductive procedure to construct the attractor is hidden inside this rule. This geometric procedure consists of a hierarchy of nested basic sets, and the attractor is just the intersection of all these sets. P.A.P. Moran was the first who had used in 1940 a geometric construction to calculate the Hausdorff dimension of a Cantor-like set. The fact is that if one knows rates of contraction of basic sets during their transformation from the previous generation to the next one, and if one knows how many basic sets of the next generation are inside the ones of the previous generation and how they are nested inside each other, then one can calculate the Hausdorff dimension of the attractor. The calculation is expressed in the form of the Bowen's equation; R. Bowen had derived this equation in a particular situation. Thus, one can be convinced that there is a well developed machinery allowing one to calculate or to estimate the Hausdorff dimension of a fractal set. There are many deep results related to Hausdorff and box dimensions, see [37,55,59,61,97,111,115] and references therein.

Unfortunately not always geometric and metrical features of invariant sets reflect dynamics occurring on them. The simplest example is the 2-dimensional torus: it can serve as the phase space for an Anosov map and for a system with quasi-periodic orbits as well. So, one needs characteristics reflecting not only a geometry but also a dynamics. And our book is devoted to a description of one of such characteristics – the spectrum of dimensions for Poincaré recurrences.

Poincaré recurrences are the main indicators and characteristics of how a certain state of a dynamical system repeats itself in time. In general, the complete analysis of the return behavior of any initial condition is not feasible. A traditional approach is to study statistical properties of Poincaré recurrences, i.e., to deal with typical orbits with respect to some invariant measures (see, for instance [17,19, 123] and references therein). These investigations led to a series of very interesting results. But they have the disadvantage that one does not get control on the sets of zero measure. As it was shown in [18], the remaining zero-measure set can be very large in terms of topological entropy or dimension. Moreover, the existence of many different ergodic measures indicates the fact that recurrence properties may vary with the measure. Hence, one can expect a certain multifractal nature.

Recently, a new approach has been proposed ([9,11,95]) that makes use of ideas and methods of the dimension theory ([97]). As a part of this new approach the spectrum of dimensions for Poincaré recurrences was introduced in [10] and [11]. One of the main reasons for this approach was to catch the return time properties of the entire invariant set at once. For this, ideas of the general concept of the multifractal analysis [16] were applied. The approach helps to detect the regions inside the invariant set which exhibit a certain recurrence behavior. Finally, the invariant set can be decomposed with respect to the local recurrences.

Up to now, the main objects of multifractal analysis were the multifractal decompositions associated to the pointwise dimension, Lyapunov exponents or the local entropy. The spectrum for Poincaré recurrences provides a new possibility to study multifractal features of invariant sets in dynamical systems. One of the conjectures in [16], see also [101], is that a "nice" dynamical system can have only a finite number of independent multifractal characteristics. This motivates the study of relations of the recurrence spectrum to the known spectra. In the class of systems we consider in the book we are able to show that the recurrence spectrum is determined by the spectrum of Lyapunov exponents. This result supports the conjecture and gives a new insight into the nature of recurrences: global recurrences of the system are determined by local properties along the orbits.

The recurrence spectrum is a function of a parameter, say q. This parameter tunes the balance between the distance of returns and the time needed for the return. For a given invariant set it reflects the multifractal nature of asymptotics of local return times, for example, large local instabilities favorize fast returns.

Roughly speaking, subsets with similar behavior of return times are labeled by same value of the parameter q. In particular, the corresponding outer measure sits on the set of points with a prescribed return times asymptotics, provided that it is non-zero and finite.

To be more specific, we have to remind the reader a general approach to study fractal dimensions that was proposed by Ya. Pesin in [98] and described in details in [97]. It is based on so called generalized Carathéodory construction. Let us describe it now for a simple situation when we deal with a compact metric space X, endowed with the distance $d(x, y)$. The ball $B(x, \varepsilon) = \{y\colon d(x, y) < \varepsilon\}$ of radius ε centered at x is thus well-defined. Given $Z \subset X$ one may consider a finite cover $G = \{B(x_i, \varepsilon_i) =: B_i\}$ of Z, $\bigcup_{i=1}^{N} B_i \supset Z$, $\varepsilon_i < \varepsilon$. Let $\xi : \mathcal{B} \to \mathbb{R}^+$ be a function defined on the set of all open balls in X that takes only non-negative values and goes to zero as the radius of the ball goes to zero. We define the sum

$$M_\xi(\alpha, G, Z) = \sum_{i}^{N} \xi(B_i)\varepsilon_i^\alpha$$

and consider its minimum

$$M_\xi(\alpha, \varepsilon, Z) = \inf_G \sum_{i}^{N} \xi(B_i)\varepsilon_i^\alpha,$$

taken over all finite covers of Z by balls of radius less than or equal to ε. It is clear that $M_\xi(\alpha, \varepsilon Z)$ is monotone function in ε, therefore there exists the limit

$$m(\alpha, Z) = \lim_{\varepsilon \to 0} M_\xi(\alpha, \varepsilon, Z).$$

It follows [97] that there exists a critical value $\alpha_c \in [-\infty, \infty]$ such that

$$m(\alpha, Z) = \begin{cases} 0, & \alpha > \alpha_c,\ \alpha_c \neq +\infty, \\ \infty, & \alpha < \alpha_c,\ \alpha_c \neq -\infty. \end{cases}$$

The number α_c is said to be the Carathéodory dimension of Z (see details below).

Among other features of α_c the following one is related to finding of the average value of the function ξ. Indeed, it follows directly from the definition that, if $\varepsilon_i < \varepsilon \ll 1$, then the sum

$$\sum_{i}^{N} \xi(B_i)\varepsilon_i^\alpha \quad \begin{cases} \gg 1, & \text{if } \alpha < \alpha_c, \\ \ll 1, & \text{if } \alpha > \alpha_c. \end{cases}$$

For the sake of simplicity assume that

$$\sum_{i=1}^{N} \xi(B_i)\varepsilon_i^{\alpha_c} \approx 1.$$

Then,

$$\langle \xi(B_i)\varepsilon_i^{\alpha_c} \rangle = \frac{1}{N}\sum_{i=1}^{N} \xi(B_i)\varepsilon_i^{\alpha_c} \approx \frac{1}{N}.$$

For $\varepsilon_i = \varepsilon$, the minimal number of balls of radius ε needed to cover set Z is $N \approx \varepsilon^{-b}$, where $b = \dim_B Z$ is the fractal dimension (the box dimension, if it exists) of Z. Therefore,

$$\langle \xi(B_i) \rangle \approx \varepsilon^{b-\alpha_c}$$

if $\alpha_c \neq \pm\infty$ and $b \neq \pm\infty$.

Thus, the generalized Carathéodory construction allows one to estimate the arithmetic average of functions of balls (or open sets) over subsets of X.

We apply now the construction to study Poincaré recurrences for the dynamical system (f^t, X), $t \in \mathbb{Z}$. For an open ball $B := B(x, \varepsilon)$ we define the Poincaré recurrence as

$$\tau(B) = \inf\{\tau(x, B)\colon\ x \in B\}$$

where

$$\tau(x, B) = \min\{t > 0\colon\ f^t x \in B\}$$

is the first return time of $x \in B$. Given a real non-negative function $\eta : \mathbb{R}^+ \to \mathbb{R}^+$ such that $\eta(t) \to 0$ as $t \to \infty$, let $\xi_q(B) = \eta(\tau(B))^q$, $q \geqslant 0$. Then we apply the Carathéodory construction outlined above to obtain

$$\langle \xi_q(B) \rangle \approx \varepsilon^{b-\alpha_c}$$

where the critical value α_c depends on q now: $\alpha_c = \alpha_c(q)$. The function $\alpha_c(q)$ is called the spectrum of dimensions for Poincaré recurrences. If there exists $q_0 > 0$ such that $\alpha_c(q) > 0$, $0 \leqslant q < q_0$ and $\lim_{q\to q_0} \alpha_c(q) = 0$, then

$$\langle \xi_{q_0}(B) \rangle \approx \varepsilon^{b}.$$

We call q_0 the dimension for Poincaré recurrences (in literature it is called sometimes the AP-dimension [36,76,95,96,125]).

There are two important cases we want to mention. If the dynamical system has positive topological entropy, then for the function $\eta(t) = e^{-t}$ one can find such positive q_0, i.e.,

$$\langle e^{-q_0\tau(B)} \rangle \approx e^{b}$$

and one can expect that the average Poincaré recurrence for balls of radius ε behaves as

$$\langle \tau(B) \rangle \approx -\frac{b}{q_0}\ln\varepsilon \tag{1.1}$$

but if the topological entropy equals 0, then very often positive q_0 exists for $\eta(t) = 1/t$, i.e.,

$$\langle \tau(B)^{-q_0} \rangle \approx e^b$$

and one may assume that

$$\langle \tau(B) \rangle \approx \varepsilon^{-b/q_0}. \tag{1.2}$$

The asymptotic relations (1.1) and (1.2) not only manifest the "physical meaning" of the dimension for Poincaré recurrences but also can serve as the basis for algorithms to calculate q_0 in specific situations (see Chapter 16).

Now we assume that μ is a Borel probability measure on X. As it was said, the function $\alpha_c(q)$ depends on a set Z of initial points, i.e., $\alpha_c(q) = \alpha_c(q, Z)$. We may define the spectrum for the measure μ as follows,

$$\alpha_c^\mu(q) := \inf\{\alpha_c(q, Z)\colon\ \mu(Z) = 1\}.$$

Thus, one can compare different measures with respect to their recurrence properties.

As usual, the question of the validity of variational principle arises: is it true that $\sup \alpha_c^\mu(q) = \alpha_c(q)$, where supremum is taken over all Borel probability measures? For maps acting on invariant Cantor sets resulting from Moran-type constructions, this question was positively answered in [5,7] and will be described in Chapter 5 of the book.

A local version of $\alpha_c^\mu(q)$ was introduced in [5,39] and [7] in the following way: the quantity

$$\alpha_c^\mu(q, x) = \liminf_{\varepsilon \to 0} \inf_{y \in B(x,\varepsilon)} \frac{\log \mu(B(y, \varepsilon)) + q\tau(B(y, \varepsilon))}{\log \varepsilon}$$

is called the lower q-pointwise dimension of μ at the point x. It was shown in [39] that if this local dimension equals a constant μ-almost everywhere, then the dimension for the measure μ also is equal to the constant. Such situations occur for ergodic invariant measures.

The book is organized as follows. The first part is devoted to main notions, ideas and methods we want to expose, such as symbolic dynamics, including topological pressure; geometric constructions as inductive procedures to obtain invariant sets in the phase space and definition and main properties of spectrum of dimensions for Poincaré recurrences. A reader who knows about symbolic dynamics can skip Chapter 2. Part II deals with zero-dimensional invariant sets on which the system can behave chaotically or regularly. In Part III we describe one-dimensional systems with discrete time (generated by Markov maps) and with continuous time in the form of suspended flows. Part IV deals with dimensions for invariant measures, local dimensions and the variational principle. In Part V we discuss some applications of the developed theory. For readers who are familiar with the machinery of statistical mechanics, we wrote Chapter 14, where ideas and main results of our book are explained in an intuitive way. In Appendices, Part VI, we list some necessary items for the convenience of readers.

We wish to thank our friends and collaborators, including Jean-Renne Chazottes, Ricardo Coutinho, Bastien Fernandez, Wen-Wei Lin, Alejandro Maass, Yakov Pesin, Nikolai Rulkov, Benoît Saussol, Joerg Schmeling, Victor Sirvent, Sandro Vaienti and George Zaslavsky for their invaluable and fruitful discussions and help. The first author is grateful to Ya. Pesin who attracted him to the beautiful field of fractal dimensions. We specially thank G.M. Zaslavsky without whose persuasiveness the book would never appear.

PART I

FUNDAMENTALS

Chapter 2

Symbolic Systems

In spite of the fact that dynamical systems are defined by a local rule, say a map $x \mapsto f(x)$ (it could be a system of ODE $\dot{x} = f(x)$, but we restrict ourself to the case of discrete time), and this rule is often expressed in a simple form, the global behavior of orbits could be amazingly complex. Here, a (semi-)orbit through an initial point x_0 is $\Gamma(x_0) := \bigcup_{i=0}^{\infty} f^i x_i$; a union of orbits Y is an invariant set: $f(Y) \subset Y$. Complexity of such a behavior is reflected in the geometry of invariant sets and can be measured by Hausdorff and box dimensions and other dimension-like characteristics. Invariant sets are constructed by using symbolic dynamical models.

In this chapter we remind general facts of the symbolic dynamics theory (see for instance [49]) and emphasize some of them which are not so well known.

2.1. Specified subshifts

Given $p \in \mathbb{N}$, consider the set $\Omega_p = \{0, \ldots, p-1\}^{\mathbb{Z}^+}$ endowed with the product topology which makes Ω_p a compact metrizable space ($\mathbb{Z}^+ = \mathbb{N} \cup \{0\}$). A point in Ω_p is denoted by $\omega = \{\omega_k\}_{k \geqslant 0}$. A subshift is the dynamical system (Ω, σ) where $\sigma : \Omega \to \Omega$ is the shift operator, defined by $(\sigma\omega)_k = \omega_{k+1}$, and $\Omega \subset \Omega_p$ is a closed and σ-invariant subset.

Given the word $\underline{\omega} := \omega_0 \ldots \omega_i \ldots \omega_n$, $\omega_i \in \{0, \ldots, p-1\}$, the subset

$$[\omega_0 \ldots \omega_n] := \{\omega' \in \Omega\colon\ \omega'_i = \omega_i,\ 0 \leqslant i \leqslant n\}$$

is called a cylinder. Every cylinder is open and closed. A word $\underline{\omega} = \omega_0 \ldots \omega_i \ldots \omega_n$, $\omega_i \in \{0, \ldots, p-1\}$, is said to be admissible in Ω if the corresponding cylinder $[\omega_0 \ldots \omega_n]$ is not empty. We will abuse notation and express admissibility by writing $\underline{\omega} \in \Omega$. If $\underline{\omega} = [\omega_0 \ldots \omega_{n-1}]$ then $|\underline{\omega}| := n$.

A subshift (Ω, σ) is said to be (topologically) mixing if for any two admissible words $\omega_0 \ldots \omega_n$ and $\omega'_0 \ldots \omega'_m$ there exists $k_0 \in \mathbb{Z}^+$ (the mixing time) such that for any $k \geqslant k_0$, there exists a word $\omega_0'' \ldots \omega_k''$ such that the concatenated word

$$\omega_0 \ldots \omega_n \, \omega_0'' \ldots \omega_k'' \, \omega_0' \ldots \omega_m'$$

is admissible in Ω.

The subshift (Ω, σ) is said to be specified (or to have the specification property) if there exists $n_0 \in \mathbb{Z}^+$ such that for any pair of admissible words, $\omega_0 \ldots \omega_n$ and $\omega_0' \ldots \omega_m'$, there exists $k \leqslant n_0$ and a word $\omega_0'' \ldots \omega_k''$ such that the concatenated word $\omega_0 \ldots \omega_n \, \omega_0'' \ldots \omega_k'' \, \omega_0' \ldots \omega_m'$ is admissible in Ω.

We now define a class of distance functions on Ω consistent with the product topology. Given $a \in \{0, \ldots, p-1\}$ let

$$[a] = \{x \in \Omega\colon \, x_0 = a\} \quad \text{and} \quad \zeta = \{[a]\colon \, a = 0, \ldots, p-1\}$$

denote the partition of Ω into 1-cylinders. Denote by ζ^n the dynamical partition $\zeta^n := \bigvee_{j=0}^{n-1} \sigma^{-j}\zeta$. Then $\zeta^n(\omega)$ will be the atom of the refined partition ζ^n that contains ω and will be referred to as the n-cylinder about ω. Given a continuous function $u : \Omega \to (0, \infty)$ we endow Ω with the metric d_Ω defined by $d_\Omega(\omega, \omega') := \mathrm{e}^{-u(\zeta^n(\omega))}$ where n is such that $\omega' \in \zeta^n(\omega)$ and $\omega' \notin \zeta^{n+1}(\omega)$, and

$$u\big(\zeta^n(\omega)\big) = \sup_{k \leqslant n} \sup_{\omega' \in \zeta^k(\omega)} \big(u(\omega') + u(\sigma\omega') + \cdots + u\big(\sigma^{k-1}\omega'\big)\big),$$
$$n = 1, 2, \ldots.$$

Remark that the standard metric is recovered when one chooses $u \equiv 1$. If one chooses $u(\omega) = -\log \lambda(\omega_0)$, which is a constant on every atom of ξ, then

$$d_\Omega(\omega, \omega') = \prod_{\ell=0}^{n-1} \lambda(\omega_\ell), \quad \text{and} \quad \operatorname{diam} \xi_n(\omega) = \prod_{\ell=0}^{n-1} \lambda(\omega_\ell), \tag{2.1}$$

i.e., we have a situation similar to that encountered in Moran-like geometric constructions, where $0 < \lambda(\omega_0) < 1$, see Section 3.

LEMMA 2.1. *d_Ω is a distance function on Ω.*

PROOF. Let us prove it for the case (2.1). It is clear that $d_\Omega(\omega, \omega') = 0$ if and only if $\omega = \omega'$. Then, we show the triangle inequality

$$d_\Omega(\omega, \omega') \leqslant d_\Omega(\omega, \omega'') + d_\Omega(\omega'', \omega'), \tag{2.2}$$

with integers i, j and l such that $\omega_k = \omega_k'$ for $k < i$ and $\omega_i \neq \omega_i'$, $\omega_k = \omega_k''$ for $k < j$ and $\omega_j \neq \omega_j''$, and $\omega_k' = \omega_k''$ for $k < l$ and $\omega_l' \neq \omega_l''$. To show (2.2) we need to verify the inequality

$$\lambda_{\omega_0} \cdots \lambda_{\omega_{i-1}} \leqslant \lambda_{\omega_0} \cdots \lambda_{\omega_{j-1}} + \lambda_{\omega_0'} \cdots \lambda_{\omega_{l-1}'}$$

for which we have the following cases.

(a) For $0 < j \leqslant i$, it follows immediately that $\lambda_{\omega_0} \cdots \lambda_{\omega_{i-1}} \leqslant \lambda_{\omega_0} \cdots \lambda_{\omega_{j-1}}$.
(b) For $j > i > 0$, one has $l = i$. Hence, $\lambda_{\omega_0} \cdots \lambda_{\omega_{i-1}} \leqslant \lambda_{\omega_0} \cdots \lambda_{\omega_{j-1}} + \lambda_{\omega_0} \cdots \lambda_{\omega_{i-1}}$.

When $\omega_0 \neq \omega'_0$, one has that $1 \leqslant 1 + \lambda_{\omega'_0} \cdots \lambda_{\omega'_{l-1}}$ or $1 \leqslant 1 + \lambda_{\omega_0} \cdots \lambda_{\omega_{i-1}}$ in the "worst" case. □

In a similar way one proves that d_Ω is a distance in a more general situation where u depends on more symbols. Moreover, if one chooses a Hölder continuous function u, then one gets the distance used to generate Cantor-like sets in $\mathbb{R}^d$ modeled by subshifts, see Section 3.1.1.

Given $\omega \in \Omega$ and $\varepsilon \geqslant 0$ we denote by $B(\omega, \varepsilon)$ the open ball of radius ε centered at ω.

2.1.1. Ultrametric space

The following statement [5] is useful for further considerations and have an independent pedagogical interest.

STATEMENT. *The space (Ω, d_Ω) is ultra-metric, i.e.,*

$$d_\Omega(\omega, \omega') \leqslant \max\{d_\Omega(\omega, \varpi), d_\Omega(\varpi, \omega')\}$$

for every ϖ. Furthermore, for any $\omega \in \Omega$ and $\varepsilon > 0$ we have

1. *$B(\omega, \varepsilon) = \xi_{n_{\omega,\varepsilon}}(\omega)$, where*
$$n_{\omega,\varepsilon} = \min\{n \in \mathbb{N}\colon \mathrm{e}^{-u(\xi_n(\omega))} < \varepsilon\}.$$
2. *The topology generated by d_Ω is equivalent to the product topology.*

PROOF. Let $\omega, \omega' \in \Omega$, with $\omega \neq \omega'$. There exists n such that $\xi_n(\omega) = \xi_n(\omega')$ but $\xi_{n+1}(\omega) \neq \xi_{n+1}(\omega')$. This implies that

$$d_\Omega(\omega, \omega') = \mathrm{e}^{-u(\xi_n(\omega))} = \mathrm{e}^{-u(\xi_n(\omega'))}.$$

For any $\varpi \in \Omega$ either $\varpi \notin \xi_{n+1}(\omega)$ or $\varpi \notin \xi_{n+1}(\omega')$. Suppose for simplicity that $\varpi \notin \xi_{n+1}(\omega)$. Then there exists $k \leqslant n$ such that $\varpi \in \xi_k(\omega)$ but $\varpi \notin \xi_{k+1}(\omega)$, hence $d_\Omega(\omega, \varpi) = \mathrm{e}^{-u(\xi_k(\omega))}$. Since $u(\xi_k(\omega))$ is increasing, we get that $d_\Omega(\omega, \varpi) \geqslant d_\Omega(\omega, \omega')$. This proves that

$$d_\Omega(\omega, \omega') \leqslant \max\{d_\Omega(\omega, \varpi), d_\Omega(\varpi, \omega')\}.$$

Thus, d_Ω is a distance, and in addition the space (Ω, d_Ω) is ultra-metric.

Let $\omega \in \Omega$. Let $\varepsilon > 0$, and set

$$n_\varepsilon = \min\{n \in \mathbb{N}\colon \mathrm{e}^{-u(\xi_n(\omega))} < \varepsilon\}.$$

For any $\varpi \in \xi_{n_\varepsilon}(\omega)$, $\varpi \neq \omega$, there exists $n \geqslant n_\varepsilon$ such that $\varpi \in \xi_n(\omega)$ but $\varpi \notin \xi_{n+1}(\omega)$, and by definition

$$d_\Omega(\omega, \varpi) = \mathrm{e}^{-u(\xi_n(\omega))} \leqslant \mathrm{e}^{-u(\xi_{n_\varepsilon}(\omega))} < \varepsilon.$$

Thus $\varpi \in B(\omega, \varepsilon)$.

Let $\varpi \in B(\omega, \varepsilon)$, $\varpi \neq \omega$, and n such that $\varpi \in \xi_n(\omega)$ but $\varpi \notin \xi_{n+1}(\omega)$. By definition we have

$$\mathrm{e}^{-u(\xi_n(\omega))} = d_\Omega(\omega, \varpi) < \varepsilon,$$

hence, $n \geqslant n_\varepsilon$, and $\varpi \in \xi_{n_\varepsilon}(\omega)$. This proves that any ball is indeed a cylinder, and Statement 2 is now an immediate consequence. □

By using the metric d_Ω we avoid many difficulties related to comparing covers by balls and by cylinder.

2.2. Ordered topological Markov chains

A set $\Omega_A \subset \Omega_p$ is defined by a transition matrix A by declaring that $\omega \in \Omega_A$ iff $A_{\omega_k,\omega_{k+1}} = 1$, $k \geqslant 0$. Any mixing topological Markov chain has the specification property and positive topological entropy. More detailed information about Markov chains can be found for instance in [49].

In dealing with Markov maps of the interval the ordered nature of Cantor subsets of the interval will be important. This order is equivalent, through conjugacy, to a complete order of the set Ω_A that we introduce as follows. First we accept the order $0 < 1 < \cdots < p-1$ in the set of single symbols. A sign function

$$s : \{0, 1, \ldots, p-1\} \to \{-1, +1\}$$

is introduced (there are 2^p options for the function s) and extend it to words of finite length by the product

$$s(\omega_0, \omega_1, \ldots, \omega_{n-1}) = s(\omega_0) \cdot s(\omega_1) \cdot \cdots \cdot s(\omega_{n-1}).$$

The order relation in Ω_A is defined as follows. Let ω and $\omega' \in \Omega_A$ have least integer k such that $\omega_k \neq \omega'_k$. Then $\omega < \omega'$ if and only if, either $\omega_k < \omega_{k'}$ and $s(\omega_0, \omega_1, \ldots, \omega_{k-1}) = 1$ or $\omega_k > \omega_{k'}$ and $s(\omega_0, \omega_1, \ldots, \omega_{k-1}) = -1$. Otherwise $\omega > \omega'$. A topological Markov chain Ω_A that is endowed with a complete order by means of a sign function s is denoted by the triple (Ω_A, σ, s). When the sign function is the constant function $s = 1$, the order introduced in Ω_A coincides with the lexicographical order.

Next, we take the order in Ω_A as the starting point to construct a nested sequence

$$\Omega_n \subset \Omega_{n+1} \subset \cdots \subset \Omega_A$$

of mixing topological Markov chains, for a certain sufficiently large integer n (see below). We will use this construction to calculate dimension-like characteristics of invariant sets.

For each ω_0 and $n > 0$, let $\omega_0\hat{\omega}_1 \dots \hat{\omega}_{n-1}$ denote the least (in the order defined) admissible word in (Ω_A, s) of length n that begins with a given symbol ω_0, i.e.,

$$\omega_0\hat{\omega}_1 \dots \hat{\omega}_{n-1} \leqslant \omega_0\omega_1 \dots \omega_{n-1}$$

for every admissible word $\omega_0\omega_1 \dots \omega_{n-1}$.

LEMMA 2.2. *For each $i \in \{0, \dots, p-1\}$, the limit point in (Ω_A, σ)*

$$\hat{\omega}^i := \lim_{n\to\infty} i\,\hat{\omega}_1 \dots \hat{\omega}_{n-1}$$

is eventually periodic. The transient prefix of $\hat{\omega}^i$ and the period-defining word have together length at most $2p$. The point $\hat{\omega}^i \in [\,i\,] \subset \Omega_A$ is the only point such that $\hat{\omega}^i \leqslant \omega$ for every $\omega \in [\,i\,]$.

PROOF. From the ith row of transition matrix A we take $\hat{\omega}_1 \in \{j\colon A_{i,j} = 1\}$ such that

$$i\,\hat{\omega}_1 < ik \quad \text{for every } k \in \{j \neq \hat{\omega}_1\colon A_{i,j} = 1\}.$$

The choice of $\hat{\omega}_1$ is unique. Next, from the $\hat{\omega}_1$-th row of transition matrix A we take $\hat{\omega}_2 \in \{j\colon A_{\hat{\omega}_1,j} = 1\}$ such that

$$i\,\hat{\omega}_1\omega_2 < i\,\hat{\omega}_1 k \quad \text{for every } k \in \{j \neq \hat{\omega}_2\colon A_{\omega_1,j} = 1\}.$$

The choice of $\hat{\omega}_2$ is unique. Proceeding this way we will get the word $i\,\hat{\omega}_1 \dots \hat{\omega}_m$ such that for the first time the sign $s(i\,\hat{\omega}_1 \dots \hat{\omega}_{m-1})$ and symbol $\hat{\omega}_m$, simultaneously, appear again in the procedure. This will happen for $m \leqslant 2p$ (each one of the p symbols is associated to one of two possible signs). □

Thus, sequence $\hat{\omega}^i = a^i(c^i)^\infty$ where the admissible word a^i is the shortest transient prefix of $\hat{\omega}^i$ and word c^i is the period-defining word. They are unique and their lengths satisfy the inequality $|a^i| + |c^i| \leqslant 2p$ for each i, with $|a^i| \geqslant 0$ and $|c^i| \geqslant 1$. A direct consequence of the mixing property of (Ω_A, σ) is the following

LEMMA 2.3. *There exists $q \in \mathbb{N}$, $q \geqslant 3$, such that for each $i \in \{0, \dots, p-1\}$, there exists a word b^i such that*

(1) *the word $c^i b^i c^i$ is admissible in Ω_A,*
(2) *$c^i b^i c^i \neq (c^i)^q$, and*
(3) *$|c^i b^i c^i| = q|c^i|$.*

Let $B_n := \{\hat{\omega}_0^i \dots \hat{\omega}_n^i \colon i = 0, \dots, p-1\}$ and let k_n^i be such that $(c^i)^{k_n^i} \in \hat{\omega}_0^i \dots \hat{\omega}_{n-1}^i$ but $(c^i)^{k_n^i+1} \notin \hat{\omega}_0^i \dots \hat{\omega}_{n-1}^i$. The inequalities

$$n + 1 \geqslant |c^i| k_n^i \geqslant k_n^i$$

hold, and we will be considering sufficiently large values of n, $n > N_1 \geqslant 2pq$ say (see Lemma 2.4 below), so that each word in B_n has blocks $(c^i)^{k_n^i}$ that are longer than $(c^i)^q$, i.e., $k_n^i > q$, for every i. This will allow us to do profit of Lemma 2.3.

The set of admissible sequences in Ω_A that do not contain any of the words in B_n,

$$\widehat{\Omega}_n = \{\omega \in \Omega_A \colon \forall i \colon \omega_k \dots \omega_{k+n} \notin B_n,\ k \in \mathbb{Z}^+\},$$

defines a symbolic system $(\widehat{\Omega}_n, \sigma)$ which is a topological Markov chain. This system may not be mixing. However, Proposition 2.1 below tells us that it contains a mixing topological Markov chain Ω_n if n is sufficiently large. To identify Ω_n, for $k_n^i \geqslant q$ let

$$G_n = \left\{\omega_0 \dots \omega_n \in \Omega_A \colon \forall i \colon \left(c^i\right)^{\ell} \notin \omega_0 \dots \omega_n \text{ whenever } \ell \geqslant k_n^i - q\right\}.$$

The elements of G_n are, by definition, admissible in $\widehat{\Omega}_n$, $G_n \subset B_n^c$, and will be "linking sockets" for the sequences in the subset $\Omega_n \subset \widehat{\Omega}_n$. The next statement tells us that every element in G_n is a linking socket. For that we take $N_1 \geqslant k_0$, the mixing time of (Ω_A, σ).

LEMMA 2.4. *For all $n \geqslant N_1$ every two words $\omega_0 \dots \omega_n$ and $\omega'_0 \dots \omega'_n$ in G_n and for every $m > n$ there exists a word $\omega''_0 \dots \omega''_m$ such that the concatenated word*

$$\omega_0 \dots \omega_n \omega''_0 \dots \omega''_m \omega'_0 \dots \omega'_n$$

is admissible in $\widehat{\Omega}_n$.

PROOF. Because the system (Ω_A, σ) is mixing, for every $m > n \geqslant N_1$ there exists a word $\omega''_0 \dots \omega''_m$ such that the concatenated word

$$\omega_0 \dots \omega_n \omega''_0 \dots \omega''_m \omega'_0 \dots \omega'_n \tag{2.3}$$

is admissible in Ω_A. If the word is admissible in $\widehat{\Omega}_n$ too, there is nothing to prove. Otherwise, the word contains a segment of the length $n+1$ that is in B_n. In this case, Lemma 2.3 allows us to replace the forbidden segment by a segment of the same length $n+1$ that is admissible in $\widehat{\Omega}_n$. The procedure to make the replacement consists in repeating the following substitution rule as many times as necessary.

SUBSTITUTION RULE. Reading the word

$$\omega_0 \dots \omega_n \omega''_0 \dots \omega''_m \omega'_0 \dots \omega'_n$$

from left to right, at the first occurrence of a word in B_n, we replace the first occurrence of the segment $(c^i)^q$, not intersecting neither the word $\omega_0 \dots \omega_n$ nor $\omega'_0 \dots \omega'_n$, by the segment $c^i b^i c^i$. Since $m > n$, any word of length $n+1$ in (2.3) cannot intersect neither the prefix $\omega_0 \dots \omega_n$ nor the suffix $\omega'_0 \dots \omega'_n$. Hence, the substitution rule does not affect these words.

If the resulting word still contains a segment that is in B_n, we apply the rule again. The procedure is repeated as long as the reconstructed word contains a segment that is in B_n. There is a finite number of iterations after which the resulting word does not contain any segment in B_n. Indeed, the segment to be replaced at each new iteration is on the right side of the segment that was replaced in the previous iteration. The word resulting of applying the procedure may contain segments of the type $(c^i)^j$ with $j < q < k_n^i$ only. Therefore it is admissible in $\widehat{\Omega}_n$. □

The previous Lemma 2.4 will allow us to specify the subset Ω_n. A pair of words, $\omega_0 \dots \omega_n$ and $\omega'_0 \dots \omega'_n$, is said to be $\widehat{\Omega}_n$-connected if there exists a word $\omega''_0 \dots \omega''_m$ such that the concatenated word

$$\omega_0 \dots \omega_n \omega''_0 \dots \omega''_m \omega'_0 \dots \omega'_n$$

is admissible in $\widehat{\Omega}_n$. Let $C_n \subset B_n^c$ be the set of words of the length n such that each $\omega_0 \dots \omega_n \in C_n$ is $\widehat{\Omega}_n$-connected to some socket word in G_n and such that, for each $\omega_0 \dots \omega_n$, there exists a socket word in G_n which is $\widehat{\Omega}_n$-connected to $\omega_0 \dots \omega_n$. Remark that by definition we have that $G_n \subset C_n$.

Thus, the subset of $\widehat{\Omega}_n$ we consider is defined as follows

$$\Omega_n = \{\omega \in \Omega \colon \omega_k \dots \omega_{k+n} \in C_n,\ k \in \mathbb{Z}^+\}.$$

By definition, $\Omega_n \subset \widehat{\Omega}_n$ and the system (Ω_n, σ) is a topological Markov chain (words of length $n+1$ are its states).

PROPOSITION 2.1. *There exists $N \in \mathbb{N}$, $N \geqslant N_1$ such that for any $n \geqslant N$, the following properties hold.*

(i) *The set Ω_n is non-empty.*
(ii) $\Omega_n \subset \Omega_{n+1}$.
(iii) *The system (Ω_n, σ) is a mixing topological Markov chain (on words of length $n+1$).*
(iv) *There exists $m \in \mathbb{N}$ such that, for any $n \geqslant N$, the number of n-periodic orbits of (Ω_A, σ) not belonging to Ω_n is at most m.*

PROOF.

(i) Consider the periodic sequence $(c^i b^i)^\infty$ for some $i \in \{0, \dots, p-1\}$. By Lemma 2.3, this sequence is admissible in Ω_A. Moreover, the word c^i could appear consecutively at most $q-2$ times. Therefore if $k_n^i - q \geqslant q-2$, which happens if n is sufficiently large, then the periodic sequence $(c^i b^i)^\infty$ is in Ω_n.

(ii) Let $\omega_0 \dots \omega_n$ and $\omega_1 \dots \omega_{n+1}$ be two words in C_n. Since the definition of G_n implies the inclusion

$$\begin{aligned}&\{\omega \in \Omega\colon\ \omega_k \dots \omega_{k+n} \in G_n,\ k \in \mathbb{Z}^+\}\\ &\quad \subset \{\omega \in \Omega\colon\ \omega_k \dots \omega_{k+n+1} \in G_{n+1},\ k \in \mathbb{Z}^+\}.\end{aligned} \tag{2.4}$$

Then the word $\omega_0 \dots \omega_{n+1}$ is $\widehat{\Omega}_n$-connected to a word in G_{n+1}.

Moreover, if a word of length $n+1$ that is admissible in Ω_A does not belong to B_n, then none of its extensions of length $n+2$ belongs to B_{n+1}. Therefore $\widehat{\Omega}_n \subset \widehat{\Omega}_{n+1}$. Hence, the word $\omega_0 \dots \omega_{n+1}$ is $\widehat{\Omega}_{n+1}$-connected to a word in G_{n+1}. Similarly, there exists a word in G_{n+1} that is $\widehat{\Omega}_{n+1}$-connected to $\omega_0 \dots \omega_{n+1}$ and hence the word $\omega_0 \dots \omega_{n+1}$ belongs to C_{n+1}.

(iii) Let $\omega_0 \dots \omega_n$ and $\omega'_0 \dots \omega'_n$ be two admissible words in Ω_n. By definition, the word $\omega_0 \dots \omega_n$ can be $\widehat{\Omega}_n$-connected to a word $\varpi_0 \dots \varpi_n \in G_n$. The length of the connecting word is at most the number of words of length $n+1$, i.e. p^{n+1}. Similarly, there exists a word $\varpi'_0 \dots \varpi'_n \in G_n$, $\widehat{\Omega}_n$-connected to $\omega'_0 \dots \omega'_n$ by a word of length at most p^{n+1}. If $n \geqslant N_1$, then, by Lemma 2.4, the word $\varpi_0 \dots \varpi_n$ is $\widehat{\Omega}_n$-connected to $\varpi'_0 \dots \varpi'_n$ by a word of any length that is longer than n.

Consequently, any two admissible words in Ω_n of the length n can be connected by a word of arbitrary length larger than $2p^{n+1}+n$. It follows that the system (Ω_n, σ) is mixing.

(iv) Any word of length $n+1$ admissible in Ω_A but not admissible in Ω_n contains one of the words $(c^i)^{k_n^i - q}$. We have

$$n+1-(k_n^i - q)|c^i| \leqslant q|c^i| + |a^i|.$$

Consequently, the number of admissible words of length $n+1$ not admissible in Ω_n is not larger than

$$m := \sum_{i=0}^{p-1} (q|c^i| + |a^i|) p^{s|c^i|+|a^i|},$$

which is independent of n.

□

A further property of the sequence Ω_n of mixing Markov chains, proved in Section 2.4, is that the topological pressure in Ω_n of a Hölder continuous potential, defined on Ω_A but restricted to Ω_n, converges to the topological pressure in Ω_A as $n \to \infty$. This important fact will be used to find out fractal dimensions in many specific situations.

2.3. Multipermutative systems

An important class of symbolic systems, different from subshifts, consists of multipermutative systems. Let $\Omega = \{0, 1, \ldots, q-1\}^{\mathbb{N}_0}$ with the metric (2.1).

DEFINITION 2.1. A map $T : \Omega \to \Omega$ is said to be *multipermutative* if for every $\omega \in \Omega$ the sequence $T\omega$ is given by

$$T\omega = \big(\omega_0 + p_0,\ \omega_1 + p_1(\omega_0), \ldots, \omega_i + p_i(\omega_0, \ldots, \omega_{i-1}), \ldots\big)$$

with $p_0 \in A := \{0, \ldots, q-1\}$, the alphabet, and $p_i : A^i \to A$ for $i > 0$. At every coordinate the addition is understood to be modulo q.

Cylinders of length L are denoted by $\omega_L := [\omega_0, \ldots, \omega_{L-1}] \subset \Omega$, and they determine the integer value

$$\|\omega_L\|_q = \sum_{i=0}^{L-1} \omega_i q^i.$$

EXAMPLE 2.1. The *q-adic adding machine* is a multipermutative system (Ω, S) such that

$$S\omega = \big(\omega_0 + 1,\ \omega_1 + s_1(\omega_0), \ldots, \omega_i + s_i(\omega_0, \ldots, \omega_{i-1}), \ldots\big),$$

with $s_i(\omega_0, \ldots, \omega_{i-1}) = 1$ if $(\omega_0, \ldots, \omega_{i-1})$ is maximal and $s_i(\omega_0, \ldots, \omega_{i-1}) = 0$ otherwise. The word $(\omega_0, \ldots, \omega_{i-1})$ is maximal when $\omega_j = q-1$ for $j = 0, \ldots, i-1$. For every $L \geqslant 1$, a map $\{0, 1, \ldots, q^L - 1\} \to \{0, 1, \ldots, q-1\}^L$ is well-defined where $n \mapsto S^n \underline{0}_L = \omega_L$ is a bijection, and $n = \|\omega_L\|_q$. For every $L \geqslant 1$, the set $\{S^n \underline{0}_L \colon n = 0, 1, \ldots, q^L - 1\}$ is a cycle of period q^L.

The next result is a dynamical characterization of minimal multipermutative systems.

THEOREM 2.1. *For (Ω, T), a multipermutative system, the following four statements are equivalent.*

(1) *(Ω, T) is minimal.*
(2) *For every $L > 0$ and every cylinder $\omega_L := [\omega_0, \ldots, \omega_{L-1}]$, the sequence ω_L, $T\omega_L$, $T^2\omega_L, \ldots$ is periodic with smallest period q^L.*
(3) *For every $L > 0$, the numbers ("integrals") $\pi_L := \sum_{\omega_L} p_L(\omega_L)$ and the constant $p_0 \in A$ are relatively prime to q.*
(4) *The system (Ω, T) is topologically conjugate to the q-adic adding machine (Ω, S).*

PROOF. (1) $\iff$ (2). If a multipermutative map T is minimal then every point of the system must visit every cylinder set. This, together with the fact that T maps a cylinder of length L to another cylinder of the same length, implies (2). Conversely, assume that (2) holds. Then, every point of the system visits every cylinder set. Since cylinder sets span the topology of Ω, we conclude the system is minimal.

(2) $\iff$ (3). For cylinders of the length $L = 1$, the sequence $[\omega_0]$, $[\omega_0 + p_0]$, $[\omega_0 + 2p_0], \ldots$ is periodic with smallest period q if and only if $(p_0, q) = 1$. For cylinders of the length $L > 1$, assume that $\omega_L := [\omega_0, \ldots, \omega_{L-1}]$ is periodic with smallest period q^L. Then we prove that the sequence $\{T^t[\omega_0, \ldots, \omega_{L-1}, \omega_L]\}_{t=0}^{\infty}$ is periodic with smallest period q^{L+1}. It is so if and only if for every cylinder $\omega_L := [\omega_0, \ldots, \omega_{L-1}]$ we have

$$\begin{aligned} &\bigl\{\omega_L^{(k)} \in A\colon T^{kq^L}[\omega_0, \ldots, \omega_{L-1}, \omega_L] \\ &\quad = [\omega_0, \ldots, \omega_{L-1}, \omega_L^{(k)}],\ k \geqslant 0\bigr\} = A. \end{aligned} \tag{2.5}$$

A straightforward computation shows that

$$\omega_L^{(k)} = \omega_L + k\pi_L \quad \text{with } \pi_L = \sum_{\omega'_L} p_L(\omega'_L)$$

for every ω_L, where the sum is taken over all admissible words ω'_L of the length L. Indeed, define $i^{(n)} \in \{0, 1, \ldots, q-1\}$ by the equality $T^n[\omega_L, \omega_L] = [\omega_L^n, i^{(n)}]$ where $\omega_L := (\omega_0, \ldots, \omega_{L-1})$ and $i^{(0)} = \omega_L$. We want to compute $\omega_L^{(k)} = i^{(kq^L)}$, $k > 0$. For $k = 1$,

$$\begin{aligned} \omega_L^{(1)} &= i^{(q^L)} = i^{(q^L-1)} + p_L\bigl(\omega_L^{(q^L-1)}\bigr) \\ &= i^{(q^L-2)} + p_L\bigl(\omega_L^{(q^L-2)}\bigr) + p_L\bigl(\omega_L^{(q^L-1)}\bigr) \\ &= \cdots \\ &= i^{(0)} + p_L\bigl(\omega_L^{(0)}\bigr) + p_L\bigl(\omega_L^{(1)}\bigr) + \cdots + p_L\bigl(\omega_L^{(q^L-1)}\bigr), \end{aligned}$$

i.e., for every ω_L we have that $\omega_L^{(1)} = \omega_L + \pi_L$, with the integral $\pi_L = \sum_{\omega'_L} p_L(\omega'_L)$. Next, for $k > 1$, assume that $\omega_L^{(k)} = \omega_L + k\pi_L$, for every $\omega_L \in A$. Then, take ω_L to be $\omega_L^{(1)}$ to obtain

$$\omega_L^{(1)} + k\pi_L = \bigl(\omega_L^{(1)}\bigr)^{(k)} = \omega_L + (k+1)\pi_L$$

and, by definition, $(\omega_L^{(1)})^{(k)} = \omega_L^{(k+1)}$. Hence, condition (2.5) is satisfied whenever $(\pi_L, q) = 1$. This proves (2) $\iff$ (3).

(2) $\iff$ (4). The fact that (4) implies (2) is the direct corollary of the definition of the adding machine. Every cylinder $[\omega_0, \ldots, \omega_{L-1}]$ of the length L

is the image of the cylinder $[0, \ldots, 0]$ by a map T^t, where the smallest positive such t is uniquely defined by coordinates $\omega_0, \ldots, \omega_{L-1}$. Thus, for every $L > 1$ and any cylinder set $[\omega_0, \ldots, \omega_{L-1}]$, we can find out a unique sequence $(k_0, \ldots, k_{L-1}) \in \{0, \ldots, q-1\}^L$ such that

$$T^t[0, \ldots, 0] = [\omega_0, \ldots, \omega_{L-1}] \quad \text{and} \quad t = \sum_{n=0}^{L-1} k_n q^n.$$

We put $\varphi([\omega_0, \ldots, \omega_{L-1}]) := [k_0, \ldots, k_{L-1}]$ and let $\varphi(\omega) := (k_0, k_1, \ldots)$ such that for every $L > 0$, $\varphi([\omega_0, \ldots, \omega_{L-1}]) = [k_0, \ldots, k_{L-1}]$. The map $\varphi : \Omega \to \Omega$ is one-to-one by construction. Moreover, φ is continuous, since

$$[w_0, \ldots, w_{l-1}] = \varphi^{-1}(k_0, k_1, \ldots, k_{L-1}).$$

Thus, $\varphi^{-1} \circ S \circ \varphi = T$, which proves that T and S are topologically conjugate. □

The next section is devoted to a more general situation.

2.3.1. *Polysymbolic generalization*

In this section we consider multipermutative systems with different alphabets $A_i = \{0, \ldots, q_i - 1\}$ at every coordinate $i = 0, 1, \ldots$. Such systems will be referred to as polysymbolic systems. The set of sequences is $\Omega_{q_*} = A_0 \times A_1 \times \cdots$, and the size of alphabets at every coordinate i is denoted by the sequence $q_* = (q_0, q_1, \ldots)$ of positive integers $q_i = \#(A_i)$, $i \geqslant 0$. A polysymbolic system (Ω_{q_*}, T) is multipermutative if for every $\omega = (\omega_0, \omega_1, \ldots)$ the map $T : \Omega_{q_*} \to \Omega_{q_*}$ is defined by

$$T\omega = \big(\omega_0 + p_0,\ \omega_1 + p_1(\omega_0), \ldots, \omega_i + p_i(\omega_0, \ldots, \omega_{i-1}), \ldots\big)$$

with $p_i : A_0 \times \cdots \times A_{i-1} \to A_i$ and $p_0 \in A_0$.

EXAMPLE 2.2. A polyadic adding machine (Ω_{q_*}, S) (called an odometer, too) is defined as the usual q-adic adding machine, except that a word

$$(\omega_0, \ldots, \omega_{i-1}) \in A_0 \times \cdots \times A_{i-1}$$

is maximal when $\omega_j = q_j - 1$ for each $j = 0, \ldots, i-1$.

Theorem 2.1 is extended to polysymbolic systems as follows.

THEOREM 2.2. *Let (Ω_{q_*}, T) be a multipermutative system which is polysymbolic. Then the following statements are equivalent.*

(P1) *(Ω_{q_*}, T) is minimal.*

(P2) *For every* $L > 0$ *and every cylinder* $\omega_L := [\omega_0, \dots, \omega_{L-1}]$ *the sequence* $\omega_L, T\omega_L, T^2\omega_L, \dots$ *is periodic with smallest period* $\prod_{i=0}^{L-1} q_i$.

(P3) *For every* $L > 0$ *the numbers* $\pi_L := \sum_{\omega_L} p_L(\omega_L)$ *and the constant* $p_0 \in A_0$ *are relatively prime to* q_L *and* q_0, *respectively.*

(P4) (Ω_{q_*}, T) *is topologically conjugate to the polyadic adding machine* (Ω_{q_*}, S).

The proof is quite similar to the proof of Theorem 2.1. The only remark is that the polyadic expansion of an integer $t \geqslant 0$ is defined by the formula

$$t = k_0 + \sum_{i=1}^{\infty} k_i \left(\prod_{j=0}^{i-1} q_j \right) = (k_0, k_1, \dots) \tag{2.6}$$

with $k_i \in \{0, \dots, q_i - 1\}$. With the polyadic expansion of $t \geqslant 0$, the labeling of cylinders is formally identical to the one used in the proof of Theorem 2.1. The labeling of cylinders provides the coding φ for the polysymbolic case.

2.3.2. *Topological conjugation of polysymbolic minimal systems*

Theorems 2.1 and 2.2 tell us that multipermutative systems can be seen as adding machines. In this section we give the main dynamical properties multipermutative systems inherit from adding machines.

A direct consequence of Theorem 2.2 is that minimal multipermutative systems are uniquely ergodic. A proof of the following statement can be found in [120].

PROPOSITION 2.2. *Every minimal multipermutative system*

$$(T, \Omega_{q_*}), \quad q_* = (q_0, q_1, \dots), \ q_i \in \mathbb{Z}^+,$$

is uniquely ergodic. The uniquely ergodic measure μ_{q_*} *is defined over cylinder sets* $[w_0, \dots, w_{L-1}]$ *by*

$$\mu_{q_*}([w_0, \dots, w_{L-1}]) = \lim_{N \to \infty} \frac{1}{N} \sum_{i=0}^{N-1} \mathbf{1}_{[w_0, \dots, w_{L-1}]}(x) = \frac{1}{q_0 \cdots q_{L-1}},$$

where $x \in \Omega_{q_*}$ *is arbitrary.*

Topological conjugacy of multipermutative systems with adding machines allows one to determine a complete system of topological invariants. These invariants are, in fact, eigenvalues in the topological discrete spectrum.

Let us remind some definitions. Let f be a complex-valued continuous function on Ω_{q_*} which is not identically 0. A function f is an eigenfunction for $T : \Omega_{q_*} \to$

Ω_{q*} if there exists a complex number λ such that $f(T\omega) = \lambda f(\omega)$, $\forall\omega \in \Omega_{q*}$. We say, as in Definition 5.9 in [120], that T has the topological discrete spectrum if the smallest closed linear subspace of $C(\Omega_{q*})$ containing all the eigenfunctions of T is $C(\Omega_{q*})$, i.e., the eigenfunctions provide a basis for $C(\Omega_{q*})$.

It is known (see, for instance, Theorem 3.4 and Theorem 5.19 in the book [120]) that topological conjugacy for adding machines is a spectral property because they have topological discrete spectrum. The following result is a direct consequence of Theorem 5.19 in [120].

THEOREM 2.3. *Two minimal multipermutative systems are topologically conjugate if and only if they have the same eigenvalues.*

The eigenvalues for multipermutative systems can be easily computed. To simplify bookkeeping, denote by

$$\overline{[w_0, \ldots, w_{L-1}]} := k_0 + \sum_{i=1}^{\infty} k_i \left(\prod_{j=0}^{i-1} q_j \right),$$

where $[k_0, \ldots, k_{L-1}] = \varphi([w_0, \ldots, w_{L-1}])$ (see the proof of Theorem 2.1). It follows that

$$\overline{T[w_0, \ldots, w_{L-1}]} = 1 + \overline{[w_0, \ldots, w_{L-1}]}.$$

LEMMA 2.5. *Let T be a minimal multipermutative system that is specified by a sequence $q_* = (q_0, q_1, \ldots)$. Then, T has topological discrete spectrum with eigenvalues*

$$\lambda_{k,L} = \exp\left(2\pi i \frac{k}{q_0 \cdots q_{L-1}} \right)$$

and corresponding eigenfunctions

$$f_{k,L}(\omega) = \exp\left(2\pi i \frac{k}{q_0 \cdots q_{L-1}} \overline{[w_0, \ldots, w_{L-1}]} \right),$$

where $L \geqslant 1$, $0 \leqslant k < q_0 \cdots q_{L-1}$.

The proof consists of a direct verification of formulas above.

The eigenfunctions $f_{k,L}$ are piecewise constant on cylinders of length L and any other piecewise constant function is a linear combination of the eigenfunctions. The set of all piecewise constant functions is dense in $C(\Omega_{q*})$.

Sufficient and necessary conditions for the topological conjugacy of multipermutative systems can be expressed in terms of some arithmetic properties of the integer sequence q_*.

For any sequence $q_* = (q_i\colon i \geqslant 0)$ and a prime number p, we define $\#(p, q_*)$ as the number of occurrences of factor p in all elements of q_* (it could be infinity, of course). That is, $\#(p, q_*) = \sum_{i\geqslant 0} s_i$ where $s_i \geqslant 0$ is the greatest integer number such that p^{s_i} divides q_i. In the same way we define $\#(p, q_0 \dots q_n)$. This is the motivation for the following

DEFINITION 2.2. Two sequences $q_* = (q_i\colon i \geqslant 0)$ and $q'_* = (q'_i\colon i \geqslant 0)$ are *equivalent* if for every prime number p, $\#(p, q_*) = \#(p, q'_*)$.

EXAMPLE 2.3. The following sequences are equivalent: (6, 6, 6, ...), (2, 3, 3, 2, 2, 3, 2, 3, ...), (2, 3, 6, 2, 3, 6, ...), (4, 9, 4, 9, 4, ...), (12, 108, 12, 108, ...), etc.

THEOREM 2.4. *Two minimal multipermutative systems, (Ω_{q_*}, T) and $(\Omega_{q'_*}, T')$, are topologically conjugate if and only if q_* and q'_* are equivalent.*

PROOF. In view of Theorem 2.3, we prove that the systems have the same eigenvalues if and only if q_* and q'_* are equivalent. Assume the integer sequences q_* and q'_* are equivalent. Then, for every $n \geqslant 1$, $0 \leqslant k < q_0 \cdots q_{n-1}$ we have to find $m \geqslant 1$, $0 \leqslant k' < q'_0 \cdots q'_{m-1}$, such that

$$\frac{k}{q_0 \cdots q_{n-1}} = \frac{k'}{q'_0 \cdots q'_{m-1}}.$$

So, we need to prove that there exists an m such that

$$k' = k\frac{q'_0 \cdots q'_{m-1}}{q_0 \cdots q_{n-1}}. \tag{2.7}$$

is an integer. By assumption, q_* and q'_* are equivalent. Then, there exists an integer number m such that for every prime number p,

$$\#(p, q_0 \cdots q_{n-1}) \leqslant \#(p, q'_0 \cdots q'_{m-1}).$$

Thus, k' is an integer. Replacing q_* by q'_*, we conclude that the two systems have the same eigenvalues. Finally, assume that the systems have the same eigenvalues such that (2.7) holds for arbitrary n. Then, for every prime number p, one has $\#(p, q_0 \cdots q_{n-1}) \leqslant \#(p, q'_0 \cdots q'_{m-1})$. Replacing q_* by q'_* in (2.7), we obtain that for any $m > 1$ there is $n > 1$ such that

$$\#(p, q_0 \cdots q_{n-1}) \geqslant \#(p, q'_0 \cdots q'_{m-1}).$$

This proves that q_* and q'_* are equivalent. □

The theorem implies that the set of equivalence classes of sequences of positive integers is in a bijective correspondence with the classes of topologically conjugated minimal multipermutative systems.

REMARK 2.1. The technique of amalgamation and fragmentation of symbols allows one to construct homeomorphisms between minimal polyadic adding machines, see Chapter 20.

EXAMPLE 2.4. A minimal multipermutative system with a periodic distribution of alphabet sizes $q_* = (q_0, \ldots, q_{n-1}, q_0, \ldots, q_{n-1}, \ldots)$ is topologically conjugate to a q-adic adding machine with $q = \prod_{i=0}^{n-1} q_i$.

EXAMPLE 2.5. Let a q-adic adding machine and a q'-adic adding machine be such that there exist positive integers m and n for which $q^m = q'^n =: Q$. Each of them can then be transformed by amalgamation to a Q-adic adding machine, so they are topologically conjugate.

A polyadic adding machine whose alphabet size at every coordinate follows a sequence $(q_0, q_1, \ldots)$ which at every coordinate $i \geqslant 0$ has q_i a prime number is called simple.

COROLLARY 2.1. *Every minimal multipermutative system is topologically conjugate to a simple polyadic adding machine.*

2.3.3. Nonminimal multipermutative systems

In this section we consider multipermutative systems (Ω_{q_*}, T), $q_* = (q_0, q_1, \ldots)$, with multiple cycles at every cylinder of the length L.

Every multipermutative map $T : \Omega_{q_*} \to \Omega_{q_*}$ maps a cylinder of length $L \geqslant 1$ to another cylinder of the same length. Hence, at every length L the dynamics of cylinders consists eventually in a collection of cycles. We will deal with nonwandering sets such that every cylinder belongs to a cycle, and the set of all cylinders of a given length $L \geqslant 1$ is partitioned into those cycles.

The number of cycles of cylinders of the length L is denoted by $N_L \geqslant 1$, and every cycle is labeled by a symbol from the set $B_L := \{0, \ldots, N_L - 1\}$. To every multipermutative system, (Ω_{q_*}, T), we associate the sequence of positive integers $(N_1, N_2, \ldots, N_L, \ldots)$ to denote the *cycle multiplicities* at every length L. Cycle multiplicities satisfy the estimates

$$1 \leqslant N_L \leqslant \prod_{i=0}^{L-1} q_i.$$

The period of a cycle $c \in B_L$, denoted by $\tau_L(c)$, is the smallest integer such that for $\omega_L \in c$, $T^{\tau_L(c)}\omega_L = \omega_L$. The period of each cycle $c \in B_L$ satisfies the inequalities

$$1 \leqslant \tau_L(c) \leqslant \prod_{i=1}^{L-1} q_i.$$

Since we are assuming that every cylinder belongs to a cycle, then for every L,

$$\sum_{c \in B_L} \tau_L(c) = q^L.$$

Cycles are disjoint, $\forall L \geqslant 1, \forall c \neq c' \in B_L\colon c \cap c' = \emptyset$. Cycles are invariant, $\forall L \geqslant 1, \forall c \in B_L\colon Tc = c$.

We say that a cycle $c' \in B_{L+1}$ is a successor of the cycle $c \in B_L$ if there exists a cylinder $[\omega_0, \dots, \omega_{L-1}, \omega_L] \in c'$ such that $[\omega_0, \dots, \omega_{L-1}] \in c$. Denote this relation by $c < c'$.

PROPOSITION 2.3. *Every cycle has at least one successor.*

PROOF. Given $c \in B_L$ consider a cylinder $[\omega_0, \dots, \omega_{L-1}] \in c$. Then any cylinder $[\omega_0, \dots, \omega_{L-1}, \omega_L]$ determines a cycle c' such that $c < c'$. □

The set of successors of a cycle $c \in B_L$ is denoted by

$$S_{L,c} = \{c' \in B_{L+1}\colon c < c'\}.$$

The cardinality $\Gamma(c) := \#(S_{L,c})$ is said to be the branching ratio of the cycle $c \in B_L$. It is simple to see that for all $L > 0$ and every $c \in B_L$ the following inequalities hold:

$$1 \leqslant \Gamma(c) \leqslant q_{L+1} \quad \text{and} \quad \tau_L(c) \leqslant \tau_{L+1}(c') \leqslant q_{L+1}\tau_L(c).$$

To describe "branching procedure" more precisely, we need the following definitions.

DEFINITION 2.3. The number $\pi_L(c) = \sum_{\omega_L \in c} p_L(\omega_L)$, is said to be the integral associated to the cycle $c \in B_L$. For every integral $\pi_L(c)$, define a collection of integers

$$\langle \pi_L(c) \rangle := \{k\pi_L(c) \bmod q_L\colon k = 0, 1, \dots, q_L - 1\}.$$

The cardinality $\overline{\pi_L(c)} := \#(\langle \pi_L(c) \rangle)$ is said to be the order of the integral $\pi_L(c)$.

For instance, the order of the integral $0 \in A$ is $\bar{0} = 1$.

PROPOSITION 2.4. *Let $c \in B_L$ be a cycle of period $\tau_L(c)$. For every $\omega_L \in c$, the sequence of cylinders of length $L+1$*

$$[\omega_L, 0],\quad T[\omega_L, 0],\quad \ldots,\quad T^j[\omega_L, 0],\quad \ldots \tag{2.8}$$

is periodic with smallest period $\overline{\pi_L(c)}\tau_L(c)$.

PROOF. Write the sequence (2.8) in the form of a matrix with the number of columns equal to $\tau_L(c)$:

$$\begin{array}{llll}
[\omega_L, 0], & [T\omega_L, p_L(\omega_L)], & \ldots, & [T^{\tau_L(c)-1}\omega_L, \sum_{n=0}^{\tau_L(c)-2} p_L(T^n\omega_L)], \\
[\omega_L, \pi_L(c)], & [T\omega_L, \pi_L(c)+p_L(\omega_L)], & \ldots, & [T^{\tau_L(c)-1}\omega_L, \pi_L(c)+\sum_{n=0}^{\tau_L(c)-2} p_L(T^n\omega_L)], \\
\vdots & \vdots & \vdots & \vdots \\
[\omega_L, k\pi_L(c)], & [T\omega_L, k\pi_L(c)+p_L(\omega_L)], & \ldots, & [T^{\tau_L(c)-1}\omega_L, k\pi_L(c)+\sum_{n=0}^{\tau_L(c)-2} p_L(T^n\omega_L)], \\
\vdots & \vdots & \vdots & \vdots
\end{array}$$

Here we have used the fact that $\omega_L \in c$ has period $\tau_L(c)$. The first row will appear again in the matrix for the first time at the kth moment if k is the smallest integer such that $k\pi_L(c) = 0 \pmod{q_L}$, i.e., when $k = \overline{\pi_L(c)}$. □

An immediate consequence of this is the following

COROLLARY 2.2. *Let $c \in B_L$ and $c' \in B_{L+1}$ be cycles of periods $\tau_L(c)$ and $\tau_{L+1}(c')$, respectively. Then $c < c'$ if and only if $\tau_{L+1}(c') = \overline{\pi_L(c)}\tau_L(c)$.*

Hence, every cycle $c' \in B_{L+1}$ that is a successor of cycle $c \in B_L$, has the period $\tau_{L+1}(c') = \overline{\pi_L(c)}\tau_L(c)$, independent of c', and the branching ratio for cycle c is $\Gamma(c) := q_L/\overline{\pi_L(c)}$.

DEFINITION 2.4. A cycling sequence is defined to be $\xi := (c_0, c_1, \ldots, c_i, \ldots)$ such that for every $i \geqslant 0$, $c_i \in B_{i+1}$ and $c_i < c_{i+1}$. The set of all cycling sequences is denoted by

$$\mathcal{C} := \{\xi \in B_1 \times B_2 \times \cdots : c_i < c_{i+1},\ i \geqslant 0\}.$$

Remark that the number of cycles may grow exponentially fast with the length of cylinders. However, the choices for the coordinates of a cycling sequence are restricted by the relation $c_i < c_{i+1}$, for all $i \geqslant 0$, and by the fact that the branching ratios at every level are never larger than q_i. To give an appropriate description of admissible cycling sequences, let us introduce the following coding of cycles.

The code is introduced inductively from one level to the next one. First, we define "admissible" symbols for the zero-th coordinate, then for the first one, etc. Start working with the zero-th coordinate. The root (the starting point of the coding procedure) is the set Ω_{q_*} that branches out to $\Gamma(\Omega_{q_*}) := q_0/\overline{p_0}$ cycles of length $L = 1$, labeled by the symbols in the set $D_\Omega := \{0, \ldots, \Gamma(\Omega_{q_*}) - 1\}$.

(a) *Root.* The symbol $\alpha_0 \in D_\Omega$ is an admissible word.

Every cycle at every length L is designated by an admissible word defined as follows.

(b) *Successors.* For every admissible word $(\alpha_0, \ldots, \alpha_{L-1})$ there exists a set of successors

$$D_{L(\alpha_0,\ldots,\alpha_{L-1})} := \{0, \ldots, \Gamma(\alpha_0, \ldots, \alpha_{L-1}) - 1\}$$

where

$$\Gamma(\alpha_0, \ldots, \alpha_{L-1}) = q_L\, \overline{\pi_L(\alpha_0, \ldots, \alpha_{L-1})}^{\,-1}$$

is the branching ratio of the cycle corresponding to the word $(\alpha_0, \ldots, \alpha_{L-1})$, and

$$\overline{\pi_L(\alpha_0, \ldots, \alpha_{L-1})} = \sum_{\omega_L \in (\alpha_0,\ldots,\alpha_{L-1})} p_L(\omega_L).$$

The relation $(\alpha_0, \ldots, \alpha_{L-1}) < (\alpha_0, \ldots, \alpha_{L-1}, \alpha_L)$ holds for every α_L in $D_{L(\alpha_0,\ldots,\alpha_{L-1})}$.

(c) *Induction.* The construction in (b) implies that for every admissible word $\underline{\alpha}$ of length L, the word $(\underline{\alpha}, \beta)$ with $\beta \in D_{L,\underline{\alpha}}$ is admissible.

Remark that $1 \leqslant \Gamma(\alpha_0, \ldots, \alpha_{L-1}) \leqslant q_L$ for every admissible word $(\alpha_0, \ldots, \alpha_{L-1})$. Moreover, the periods of cycles satisfy the rule

$$\tau_{L+1}(\alpha_0, \ldots, \alpha_{L-1}, \alpha_L) = \overline{\pi_L(\alpha_0, \ldots, \alpha_{L-1})}\tau_L(\alpha_0, \ldots, \alpha_{L-1})$$

with initial value $\tau_1(\alpha_0) = \overline{p_0}$. From now on we use the coding introduced in (a)–(c) to label the cycling sequences in the set $\mathcal{C}$.

Next, we use the introduced coding to single out subsets in Ω in the following way. Every cycling sequence $\xi = (\alpha_0, \alpha_1, \ldots, \alpha_i, \ldots) \in \mathcal{C}$ defines a set of sequences

$$\Omega_{q,\xi} := \{\omega \in \Omega_{q_*} : [w_0, \ldots, w_i] \in (\alpha_0, \ldots, \alpha_i),\ i \geqslant 0\} \subset \Omega_{q_*}.$$

Equivalently, $\Omega_{q,\xi} = \bigcap_{L=1}^{\infty} \bigcup(\alpha_0, \ldots, \alpha_{L-1})$. Let us relate the following sequence of periods of cycles

$$\tau(\xi) = (\tau_1(\alpha_0), \tau_2(\alpha_0, \alpha_1), \ldots, \tau_{i+1}(\alpha_0, \ldots, \alpha_i), \ldots)$$

to the set $\Omega_{q,\xi}$.

THEOREM 2.5. *The collection* $\mathcal{F} := \{\Omega_{q,\xi}\colon \xi \in \mathcal{C}\}$ *satisfies the following properties.*

(1) $\mathcal{F}$ *is a partition of* Ω_{q*}.
(2) *Every* $\Omega_{q,\xi} \in \mathcal{F}$ *is* T*-invariant.*
(3) *For every* $\xi \in \mathcal{C}$ *the system* $(\Omega_{q,\xi},\ T|\Omega_{q,\xi})$ *is minimal.*

PROOF.

(1) First, we show that $\Omega_{q,\xi} \cap \Omega_{q,\xi'} = \emptyset$ for every pair of cycling sequences $\xi = (\alpha_0, \alpha_1, \ldots)$ and $\xi' = (\alpha'_0, \alpha'_1, \ldots)$, $\xi \neq \xi'$, $\xi, \xi' \in \mathcal{C}$.

Let $c_k = (\alpha_0, \ldots, \alpha_{k-1})$ and $c'_k = (\alpha'_0, \ldots, \alpha'_{k-1})$ be cycles of cylinders of length k, $k \in Z^+$, and let $c_k \cap c'_k = \emptyset$ for $k = k_0, k_0 + 1, \ldots$. A point ω is in $\Omega_{q,\xi}$ if and only if there exists a sequence of cylinders $\omega_1, \omega_2, \ldots, \omega_L, \ldots$, such that for every $L \geqslant 1$, $\omega_L \in c_{L-1}$ and $\omega = \bigcap_L \omega_L$. For the cycling sequence ξ', the point $\omega \notin c'_k$ if $k \geqslant k_0$, thus $\omega \notin \Omega_{q,\xi'}$.

Next, we have to prove that for every $\omega \in \Omega_{q*}$ there exists a $\xi \in \mathcal{C}$ such that $\omega \in \Omega_{q,\xi}$. For that, consider the sequence of cylinders $[\omega_0]$, $[\omega_0, \omega_1], \ldots, [\omega_0, \ldots, \omega_{L-1}], \ldots$ which are defined by the point $\omega \in \Omega_{q*}$. By assumption, every cylinder belongs to a cycle, i.e., there exists

$$\xi = (\alpha_0, \alpha_1, \ldots, \alpha_{L-1}, \ldots) \in \mathcal{C}$$

such that for all $L \geqslant 1$, $[\omega_0, \ldots, \omega_{L-1}] \in (\alpha_0, \ldots, \alpha_{L-1})$. Hence, $\omega \in \Omega_{q,\xi}$.

(2) Show that $T\Omega_{q,\xi} \subset \Omega_{q,\xi}$. Let $\omega \in \Omega_{q,\xi}$ for $\xi = (\alpha_0, \alpha_1, \ldots)$. Then there exists a sequence of cylinders $(\omega_1, \omega_2, \ldots, \omega_L, \ldots)$ such that for every L, $\omega_L \in (\alpha_0, \ldots, \alpha_{L-1})$ and $\omega = \bigcap_L \omega_L$. The cycles are invariant sets and for every L, $T\omega_L \in (\alpha_0, \ldots, \alpha_{L-1})$, i.e., $T\omega = \bigcap_L T\omega_L \in \Omega_{q,\xi}$. The converse inclusion is proved quite similarly.

(3) For every $\xi \in \mathcal{C}$, the set $\Omega_{q,\xi}$ is invariant, and T restricted to $\Omega_{q,\xi}$ is a map of the multipermutative type. By construction,

$$\Omega_{q,\xi} = \bigcap_{L=1}^{\infty} \bigcup (\alpha_0, \ldots, \alpha_{L-1}),$$

and for every L, the set of cylinders $(\alpha_0, \ldots, \alpha_{L-1})$ is a single cycle. Hence, $(\Omega_{q,\xi}, T)$ satisfies statement (P2) in Theorem 2.2.

□

REMARK 2.2. It is simple to show that there is a distance on Ω^+_{q*} such that the multipermutative system becomes distal. It is known [58] that every distal system can be represented in the form of an union of minimal subsets. In our theorem we present a way to construct a partition for a multipermutative system.

COROLLARY 2.3. *Every multipermutative system is topologically conjugate to a collection of simple polyadic adding machines.*

EXAMPLE 2.6. *A multipermutative system consisting of uncountably many dyadic adding machines.* Consider the multipermutative system (Ω_2, T) where $\Omega_2 = \{0, 1\}^{\mathbb{N}_0}$ and the map T is defined as follows:

(a) $p_0 = 1$,
(b) $p_{2L+1}(\omega_{2L+1}) = 0$ for $L \in \mathbb{Z}^+$,
(c) $p_{2L}(\omega_{2L}) = 1$ if $\omega_{2i} = 1$ for all $i = 0, 1, \ldots, L-1$, and $p_{2L}(\omega_{2L}) = 0$ otherwise.

For this example the inductive coding gives us the following integrals for every cycle,

$$\pi_{2L+1}(0, \alpha_1, 0, \ldots, \alpha_{2L-1}, 0) = 0 \quad \text{and} \quad \pi_{2L}(0, \alpha_1, 0, \ldots, \alpha_{2L-1}) = 1,$$

that determine the hierarchy of cycles. The branching ratios are computed to be

$$\Gamma(0, \alpha_1, 0, \ldots, \alpha_{2L-1}, 0) = 2 \quad \text{and} \quad \Gamma(0, \alpha_1, 0, \ldots, \alpha_{2L-1}) = 1,$$

from where we get the following sets of symbols for the cycling sequences,

$$D_\Omega = \{0\}, \quad D_{2L-1} = \{0, 1\}, \quad \text{and} \quad D_{2L} = \{0\}$$

for $L > 0$. Hence, the set of cycling sequences for the example is

$$\mathcal{C} = \big\{(0, \alpha_1, 0, \ldots, 0, \alpha_{2L-1}, 0, \ldots)\colon \alpha_{2L-1} \in \{0, 1\},\ L > 0\big\}.$$

The periods of cycles are $\tau_L(\underline{\alpha}) = 2^{\lceil L/2 \rceil}$, $\underline{\alpha}$ is an arbitrary admissible cycling word of the length L. Remark, for this example, that for every cycling sequence $\xi \in \mathcal{C}$ the minimal subsystem $(\Omega_{2,\xi}, T)$ is topologically conjugate to the dyadic adding machine.

2.4. Topological pressure

It was R. Bowen who first made use of the notion of the topological pressure in the theory of dynamical systems [29,30].

Let us remind Bowen's definition for subshifts (the definition for arbitrary dynamical systems can be found in [73]).

Let ψ be a real-valued continuous function on a subshift Ω. Let

$$Z_n(\psi, \Omega) = \sum_{|\underline{\omega}|=n} \exp\left(\sup_{\omega \in [\underline{\omega}]} \sum_{j=0}^{|\underline{\omega}|-1} \psi(\sigma^j \omega) \right), \tag{2.9}$$

where the sum is taken over all cylinders $[\underline{\omega}] \subset \Omega$ of length $|\underline{\omega}| = n$. It is proved in [118] that the limit

$$P_\Omega(\psi) = \lim_{n\to\infty} \frac{1}{n} \log Z_n(\psi, \Omega) \tag{2.10}$$

exists. The limit is called the topological pressure of the function ψ (the potential) on Ω with respect to σ. For every constant $c \in \mathbb{R}$, the topological pressure satisfies the property

$$P_\Omega(c + \psi) = c + P_\Omega(\psi). \tag{2.11}$$

Consider the potential $\psi \equiv 0$ then $P_\Omega(0) = h_{\text{top}}(\sigma|\Omega)$, the topological entropy. Roughly speaking, the system (σ, Ω) has $e^{h_{\text{top}} n}$ different paths of temporal length n (with some accuracy), each of them "costs"

$$\exp\left(\sum_{j=0}^{|\omega|-1} \psi(\sigma^j \omega)\right) \quad \text{units,}$$

and $e^{nP_\Omega(\psi)}$ is the total price for passing through all of them.

It is known that topological pressure is independent of the metric (preserving a given topology) and is invariant under topological conjugacy [73].

EXAMPLE 2.7. Let us calculate the topological pressure in the case where $\Omega = \Omega_A$, the topological Markov chain with a $p \times p$ transition matrix A, and the function $\psi(\omega)$ depends only on the first symbol: $\psi(\omega) = \psi(\omega_0)$. In this case

$$Z_n(\psi, \Omega) = \sum_{(i_0,\dots,i_{n-1})} \exp \sum_{j=0}^{n-1} \psi(i_j) \tag{2.12}$$

where the sum is taken over all Ω_A-admissible words $(i_0, \dots, i_{n-1})$. Set $\psi(i) = \log \rho_i$, $i = 0, \dots, p-1$, then

$$Z_n(\psi, \Omega_A) = \sum_{(i_0,\dots,i_{n-1})} \prod_{k=0}^{n-1} \rho_k. \tag{2.13}$$

It is not a difficult algebraic exercise to show that

$$Z_n(\psi, \omega_A) = RB^{n-1}E^T$$

where $R = (\rho_0, \dots, \rho_{p-1})$, $E = (1, \dots, 1)$ and

$$B = A \cdot \operatorname{diag}(\rho_0, \dots, \rho_{p-1}). \tag{2.14}$$

As a corollary of formula (2.14) we obtain that $P_\Omega(\psi) = \log \lambda_0$ where λ_0 is the spectral radius of the matrix B.

EXAMPLE 2.8. Consider a mixing Markov chain (Ω_A, σ) and a nested sequence

$$\Omega_n \subset \Omega_{n+1} \subset \cdots \subset \Omega_A$$

of mixing Markov chains constructed as described in Section 2.2. The sequence approximates the system (Ω_A, σ) in the sense of the following.

PROPOSITION 2.5. *Let $\varphi : \Omega_A \to \mathbb{R}$ be a Hölder continuous function. Then, $\lim_{n\to\infty} P_{\Omega_n}(\varphi) = P_{\Omega_A}(\varphi)$.*

PROOF. Since $\Omega_n \subset \Omega_A$, we have

$$Z_k(\varphi, \Omega_n) \leqslant Z_k(\varphi, \Omega_A), \quad k \in \mathbb{Z}^+.$$

The proof consists in obtaining an upper bound for $Z_k(\varphi, \Omega_A)$ in terms of $Z_k(\varphi, \Omega_n)$.

As argued in the proof of Proposition 2.1(iv), there exists N such that for every $n \geqslant N$ and $k \geqslant n$ a word of length $k+1$ admissible in Ω_A but not admissible in Ω_n contains the word $(c^i)^j$ for some $j \geqslant k_n^i - q$ and $i \in \{0, \dots, p-1\}$. Therefore, to each word of length $k+1$ in Ω_A, there corresponds a collection $\mathcal{I}$ (which is empty for words that are admissible in Ω_n) of disjoint collections $\{\underline{r}_l, \underline{r}_l + 1, \dots, \bar{r}_l\} \subset \{0, 1, \dots, k\}$ such that

$$\omega_{\underline{r}_l} \dots \omega_{\bar{r}_l} = \left(c^i\right)^j,$$

for some $j \geqslant k_n^i - q$. Let $C_{n,k}$ be the set of all such collections and let $\mathcal{S}_\mathcal{I}$ be the set of all words of length $k+1$ with the same associated collection $\mathcal{I}$. Changing the ordering of the summation in $Z_k(\varphi, \Omega_A)$, we have

$$Z_k(\varphi, \Omega) = \sum_{\mathcal{I}\in C_{n,k}} \sum_{\omega_0\dots\omega_k \in \mathcal{S}_\mathcal{I}} \exp\left(\sup_{\omega\in[\omega_0\dots\omega_k]} \sum_{j=0}^{k} \varphi\left(\sigma^j\omega\right)\right).$$

Lemma 2.3 allows us to map each word in $\mathcal{S}_\mathcal{I}$ to a word admissible in Ω_n. Indeed, each word $\omega_{\underline{r}_l} \dots \omega_{\bar{r}_l}$ is a concatenation of copies of $(c^i)^{k_n^i - q}$ and of $(c^i)^j$ for some $j < k_n^i - q$. In each of these copies, the last occurrence of $(c^i)^q$ is replaced by $c^i b^i c^i$. This can be done since $k_n^i - q \geqslant q$ when $n \geqslant N$ and the resulting word, say $\varpi_0 \dots \varpi_k$, is admissible in Ω_n because $q \geqslant 3$. Given $\mathcal{I}$, this procedure defines a one-to-one map.

Since φ is Hölder continuous, there exists $D > 0$ and $\theta \in (0, 1)$ such that

$$\varphi(\omega) - \varphi(\omega') \leqslant D\theta^{c(\omega,\omega')},$$

where $c(\omega, \omega')$ is the length of the longest common prefix of ω and ω'.

Consider two sequences $\omega \in [\omega_0 \dots \omega_k]$ and $\varpi \in [\varpi_0 \dots \varpi_k]$, where $\varpi_0 \dots \varpi_k$ is the word obtained by the procedure described above and where

$\sigma^{k+1}\varpi = \sigma^{k+1}\omega$. We have

$$\sum_{j=0}^{k}\varphi(\sigma^j\omega) - \sum_{j=0}^{k}\varphi(\sigma^j\varpi) \leqslant D\sum_{j=0}^{k}\theta^{c(\sigma^j\omega,\sigma^j\varpi)}. \tag{2.15}$$

Let

$$m_n = \min_{i\in\{0,\dots,p-1\}} \left(k_n^i - 2q\right)\left|c^i\right|.$$

If $c(\sigma^j\omega, \sigma^j\varpi) \geqslant m_n$, then $\theta^{c(\sigma^j\omega,\sigma^j\varpi)} \leqslant \theta^{m_n}$. This happens at most $k+1$ times in the previous sum. Moreover, $c(\sigma^j\omega, \sigma^j\varpi) = \ell$ with $1 \leqslant \ell < m_n$ only if the word $\omega_j \dots \omega_{j+m_n-1}$ has been affected by the procedure. For each ℓ, this happens at most $\lfloor (k+1)/m_n \rfloor$ times when j runs from 0 to k. Similarly, $c(\sigma^j\omega, \sigma^j\varpi) = 0$ happens at most $\max_i |b_i|\lfloor (k+1)/m_n \rfloor$ times. Thus, inequality (2.15) becomes

$$\sum_{j=0}^{k}\varphi(\sigma^j\omega) - \sum_{j=0}^{k}\varphi(\sigma^j\varpi) \leqslant D\left\lfloor \frac{k+1}{m_n}\right\rfloor\left(\sum_{\ell=1}^{m_n}\theta^\ell + \max_i |b_i|\right),$$

which implies the following inequality

$$\begin{aligned} Z_k(\varphi, \Omega_A) &\leqslant \sum_{\chi\in\mathcal{C}_{n,k}} \sum_{[\varpi_0\dots\varpi_k]\subset\Omega_n} \exp\left(\sup_{\varpi\in[\varpi_0\dots\varpi_k]} \sum_{j=0}^{k}\varphi(\sigma^j\varpi)\right. \\ &\qquad \left. + D\left\lfloor \frac{k+1}{m_n}\right\rfloor\left(\sum_{\ell=1}^{m_n}\theta^\ell + \max_i |b_i|\right)\right), \\ &\leqslant \exp\left(D\left\lfloor \frac{k+1}{m_n}\right\rfloor\left(\sum_{\ell=1}^{m_n}\theta^\ell + \max_i |b_i|\right)\right)(\#\mathcal{C}_{n,k})\ Z_k(\varphi, \Omega_n), \end{aligned}$$

where we have

$$\lim_{n\to\infty}\lim_{k\to\infty} \frac{D}{k}\left\lfloor \frac{k+1}{m_n}\right\rfloor\left(\sum_{\ell=1}^{m_n}\theta^\ell + \max_i |b_i|\right) = 0.$$

The proposition then follows because of the following result. □

LEMMA 2.6. $\lim_{n\to\infty}\lim_{k\to\infty}(1/k)\log\#\mathcal{C}_{n,k} = 0$.

PROOF. Each collection $\mathcal{I}$ is encoded by a word in $\{0, 1\}^{k+1}$ as follows. We assign the symbol 0 to each symbol indexed in the word by an integer belonging to the collection $\mathcal{I}$, and the symbol 1 is assigned to every other symbol in the word.

Using this encoding, $\mathcal{C}_{n,k}$ is in one-to-one correspondence with the set

$$\mathcal{C}'_{n,k} = \bigl\{w \in \{0,1\}^{k+1}\colon \min\bigl\{\ell\colon 0^\ell \in w\bigr\} = j\bigl|c^i\bigr|,\\ j \geqslant k_n^i - q,\ j \in \{0,\dots,k-m_n+1\}\bigr\}.$$

If a word of length m_n in $\mathcal{C}'_{n,k}$ contains 1's, then these symbols have to be consecutive. Consequently, $\mathcal{C}'_{n,k}$ is a subset of

$$\mathcal{B}_{n,k} = \bigl\{w \in \{0,1\}^{k+1}\colon w_j \dots w_{j+m_n-1} \in \mathcal{A}_n\ \forall j\bigr\},$$

where

$$\mathcal{A}_n = \bigl\{0^{\ell_1}1^{\ell_2}0^{\ell_3},\ \ell_1,\ell_2,\ell_3 \in \mathbb{Z}^+,\ \ell_1+\ell_2+\ell_3 = m_n\bigr\}.$$

Let $k, k' \geqslant n$. Any element of $\mathcal{B}_{n,k+k'}$ is the concatenation of an element of $\mathcal{B}_{n,k}$ and an element of $\mathcal{B}_{n,k'}$. Hence we have

$$\#\mathcal{B}_{n,k+k'} \leqslant (\#\mathcal{B}_{n,k})(\#\mathcal{B}_{n,k'}),$$

and using a sub-additivity argument, it follows that the limit

$$\lim_{k\to\infty} \frac{1}{k} \log \#\mathcal{B}_{n,k}$$

exists. Any element of $\mathcal{B}_{n,km_n}$ is the concatenation of k elements of $\mathcal{A}_n$. Therefore, we have

$$\#\mathcal{B}_{n,km_n} \leqslant (\#\mathcal{A}_n)^k, \quad k \in \mathbb{Z}^+.$$

Moreover, we have

$$\#\mathcal{A}_n = \frac{(m_n+1)(m_n+2)}{2},$$

and then

$$\lim_{n\to\infty}\lim_{k\to\infty} \frac{1}{km_n} \log \#\mathcal{B}_{n,km_n} \leqslant \lim_{n\to\infty} \frac{1}{m_n} \log \frac{(m_n+1)(m_n+2)}{2} = 0. \qquad \square$$

2.4.1. *Dimension-like definition of topological pressure*

Following Pesin's book [97], let us proceed in the following way. For a finite or a countable cover C of Ω by cylinders of lengths greater than n and $\beta \in \mathbb{R}$ let

$$\mathcal{Z}(\beta,\psi,C,\Omega) = \sum_{[\omega]\in C} \exp\Biggl(-\beta|\underline{\omega}| + \sup_{\omega\in[\underline{\omega}]} \sum_{j=0}^{|\underline{\omega}|-1} \psi\bigl(\sigma^j\omega\bigr)\Biggr). \tag{2.16}$$

It is proved in [97] that the topological pressure $P_\Omega(\psi)$ coincides with the threshold value

$$P_\Omega(\psi) = \sup\Bigl\{\beta\colon \lim_{n\to\infty}\bigl(\inf\{\mathcal{Z}(\beta,\psi,\mathcal{C},\Omega)\colon |\mathcal{C}| \geqslant n\}\bigr) = \infty\Bigr\}. \tag{2.17}$$

EXAMPLE 2.9. In our further study we shall need some formulas for topological pressure on non-invariant sets. The following example shows how to deal with this problem in simple situations.

For the sequence of integers $p_* = (p_1, p_2, p_1, p_2, \ldots)$, $p_1 \neq p_2$, consider the non-σ-invariant set

$$\Omega_{p_*} = \mathbb{Z}_{p_1} \times \mathbb{Z}_{p_2} \times \mathbb{Z}_{p_1} \times \mathbb{Z}_{p_2} \times \cdots,$$

of the kind involved in multipermutative systems. The topological pressure of potential ψ on Ω_{p_*} with respect to the shift σ, $P_{\Omega_{p_*}}(\psi)$, is computed from Pesin's definition in (2.16) and (2.17). First, we observe that Ω_{p_*} is σ^2-invariant and introduce a recoding as follows. Define new symbols $s = (u, v) \in \mathbb{Z}_{p_1} \times \mathbb{Z}_{p_2}$. Every $\omega \in \Omega_{p_*}$ is recoded as $s = s_0 s_1 \ldots s_n \ldots$ with $s_n = (\omega_{2n}, \omega_{2n+1})$. The encoding function $s \mapsto \omega$ is denoted by h and the set of all s-sequences by

$$\widetilde{\Omega}_{p_*} = (\mathbb{Z}_{p_1} \times \mathbb{Z}_{p_2}) \times (\mathbb{Z}_{p_1} \times \mathbb{Z}_{p_2}) \times \cdots.$$

To make use of the recoding, we write the "statistical sum" (2.16) for cylinder sets $[\underline{\omega}] \subset \Omega_{p_*}$ of even length $2m$ in the following form,

$$\begin{aligned}\mathcal{Z}(\beta,\psi,\Omega_{p_*}) &= \sum_{[\underline{\omega}]\subset\Omega_{p_*}} \exp\Biggl(-\beta(2m) \\ &\qquad + \sup_{\omega\in[\underline{\omega}]} \sum_{j=0}^{m-1}\bigl(\psi\bigl(\sigma^{2j}\omega\bigr) + \psi\bigl(\sigma^{2j+1}\omega\bigr)\bigr)\Biggr) \\ &= \sum_{[\underline{s}]\subset\widetilde{\Omega}_{p_*}} \exp\Biggl(-\tilde{\beta}m + \sup_{s\in[\underline{s}]}\sum_{j=0}^{m-1}\tilde{\psi}\bigl(\sigma^j s\bigr)\Biggr)\end{aligned} \tag{2.18}$$

where $\tilde{\beta} = 2\beta$ and $\tilde{\psi}(s) = \psi(h(s)) + \psi(\sigma h(s))$. The threshold value $\tilde{\beta}_0$ of the statistical sum in (2.18) is the topological pressure of potential $\tilde{\psi}$ on the set $\widetilde{\Omega}_{p_*}$ with respect to σ (on $\widetilde{\Omega}_{p_*}$). Since $\tilde{\beta}_0 = 2\beta_0$, we have that

$$P_{\Omega_{p_*}}(\psi) = \frac{1}{2} P_{\widetilde{\Omega}_{p_*}}(\tilde{\psi}). \tag{2.19}$$

The set $\widetilde{\Omega}_{p_*}$ is σ-invariant and the pressure $P_{\widetilde{\Omega}_{p*}}(\tilde{\psi})$ is of the Bowen type. Then, we compute it by using (2.9) and (2.10),

$$
\begin{aligned}
Z_n\big(\tilde{\psi}, \widetilde{\Omega}_{p_*}\big) &= \sum_{\substack{[s]\subset\widetilde{\Omega}_{p*} \\ \|[s]\|=n}} \exp\left(\sup_{s\in[s]} \sum_{j=0}^{n-1} \tilde{\psi}\big(\sigma^j s\big)\right) \\
&= \sum_{\substack{[s]\subset\widetilde{\Omega}_{p*} \\ \|[s]\|=n}} \exp\left(\sup_{s\in[s]} \sum_{j=0}^{n-1} \psi\big(\sigma^{2j} h(s)\big)\right) \\
&\qquad \times \exp\left(\sup_{s\in[s]} \sum_{j=0}^{n-1} \psi\big(\sigma^{2j+1} h(s)\big)\right).
\end{aligned}
\tag{2.20}
$$

In the case that potential $\psi(\omega)$ is a function of the first symbol ω_0, the partition function (2.20) may be factorized,

$$
\begin{aligned}
Z_n\big(\tilde{\psi}, \widetilde{\Omega}_{p_*}\big) &= \left(\sum_{\substack{[\omega]\subset\Omega_{p_1} \\ \|[\omega]\|=n}} \exp\left(\sup_{\omega\in[\omega]} \sum_{j=0}^{n-1} \psi\big(\sigma^j \omega\big)\right)\right) \\
&\quad \times \left(\sum_{\substack{[\omega]\subset\Omega_{p_2} \\ \|[\omega]\|=n}} \exp\left(\sup_{\omega\in[\omega]} \sum_{j=0}^{n-1} \psi\big(\sigma^j \omega\big)\right)\right),
\end{aligned}
$$

and we conclude that $P_{\widetilde{\Omega}_{p*}}(\tilde{\psi}) = P_{\Omega_{p_1}}(\psi) + P_{\Omega_{p_2}}(\psi)$, where $\Omega_k = \mathbb{Z}_k^{\mathbb{N}}$. Thus, the topological pressure of ψ on the non-invariant set Ω_{p_*} is

$$
P_{\Omega_{p_*}}(\psi) = \frac{1}{2}\big(P_{\Omega_{p_1}}(\psi) + P_{\Omega_{p_2}}(\psi)\big). \qquad \square
$$

The previous example extends to periodic sequences

$$
p_* = (p_1, \ldots, p_k, p_1, \ldots, p_k, \ldots)
$$

of arbitrary period $k > 1$ as follows,

$$
P_{\Omega_{p_*}}(\psi) = \frac{1}{k} \sum_{j=1}^{k} P_{\Omega_{p_j}}(\psi); \tag{2.21}
$$

provided ψ is a function of the first symbol only.

Chapter 3

Geometric Constructions

Many invariant sets are resulting from so-called geometric constructions [97]. Let (σ, Ω), $\Omega \subset \Omega_p = \{0, \dots, p-1\}^{\mathbb{N}}$, be a subshift, a closed σ-invariant subset of the full shift with p symbols. Consider p closed subsets $\Delta_0, \dots, \Delta_{p-1} \subset \mathbb{R}^m$. For each word $\omega_0 \dots \omega_{n-1}$ admissible in Ω define *basic sets* $\Delta_{\omega_0 \dots \omega_{n-1}}$ which satisfy the following assumptions:

(A) $\Delta_{\omega_0 \dots \omega_{n-1}}$ are closed and non-empty,
(B) $\Delta_{\omega_0 \dots \omega_{n-1} j} \subset \Delta_{\omega_0 \dots \omega_{n-1}}$, $j = 0, \dots, p-1$,
(C) $\operatorname{diam} \Delta_{\omega_0 \dots \omega_{n-1}} \to 0$ as $n \to \infty$.

We can define now a nonempty set

$$F = \bigcap_{n=1}^{\infty} \bigcup_{(\omega_0 \dots \omega_{n-1})} \Delta_{\omega_0 \dots \omega_{n-1}}. \tag{3.1}$$

The closed set F becomes a Cantor set, provided that the following “separation conditions” hold

(D) $\Delta_{\omega_0 \dots \omega_{n-1}} \cap \Delta_{\varpi_0 \dots \varpi_{n-1}} \cap F = \emptyset$ whenever $\omega_0 \dots \omega_{n-1} \neq \varpi_0 \dots \varpi_{n-1}$.

The *coding map* $\chi : \Omega \to F$ is defined as follows: for any $\omega = \omega_0, \dots, \omega_{n-1}, \dots \in \Omega$, $\chi(\omega) = x$ if $x \in \bigcap \Delta_{\omega_0 \dots \omega_{n-1}}$.

If the separation conditions in (D) are satisfied, then the map $\chi \circ \sigma \circ \chi^{-1}$ generates a dynamical system on the set F. Evidently, it is topologically conjugated to (Ω, σ) by the coding map χ.

3.1. Moran constructions

The simplest constructions are of Moran type. In this case $\Omega = \Omega_p$ and basic sets satisfy the following additional axioms:

(M_1) Every basic set is the closure of its interior,
(M_2) For any n, $\operatorname{Int} \Delta_{\omega_0 \dots \omega_{n-1}} \cap \operatorname{Int} \Delta_{\varpi_0 \dots \varpi_{n-1}} = \emptyset$ if $\omega_0 \dots \omega_{n-1} \neq \varpi_0 \dots \varpi_{n-1}$,

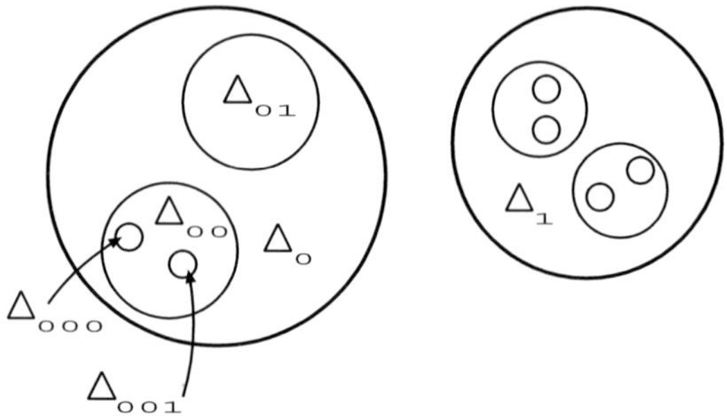

Figure 3.1. Moran construction governed by the full shift on two symbols.

(M3) The basic set $\Delta_{\omega_0\dots\omega_{n-1}j}$ is homeomorphic to $\Delta_{\omega_0\dots\omega_{n-1}}$,
(M4) There are numbers $0 < \lambda_j < 1$, $j = 0, \dots, p-1$, such that $\operatorname{diam} \Delta_{\omega_0\dots\omega_{n-1}j} = \lambda_j \operatorname{diam} \Delta_{\omega_0\dots\omega_{n-1}}$.

The basic sets of the 1st, 2nd and 3rd generations are shown on Figure 3.1.

EXAMPLE 3.1. *Let J be an invariant set of the map $g : [0, 1] \to [0, 1]$,*

$$g(x) = \begin{cases} x/\lambda_0 & \text{if } x \in [0, \lambda_0], \\ 0 & \text{if } x \in (\lambda_0, 1-\lambda_1), \\ \frac{x}{\lambda_1} - \frac{1-\lambda_1}{\lambda_1} & \text{if } x \in [1-\lambda_1, 1], \end{cases} \tag{3.2}$$

where $0 < \lambda_0 < \lambda_1 < 1$, $\lambda_0 + \lambda_1 < 1$, consisting of all points of all orbits belonging to $[0, 1]$. The set J is a Cantor set.

PROOF. The set J is constructed with the help of the contractions $u_{0,1} : [0, 1] \to [0, 1]$,

$$u_0(x) = \lambda_0 x, \quad u_1(x) = \lambda_1 x + 1 - \lambda_1,$$

such that $g \circ u_i = \text{id}$ on $[0, 1]$. For every word $\underline{\omega} = w_0 \dots w_{i-1} \in \{0, 1\}^i$ define the sets

$$\Delta_{w_0\dots w_{i-1}} := u_{w_{i-1}} \circ \dots \circ u_{w_0}\big([0, 1]\big),$$

i.e., the Δ-sets are basic sets of the geometric construction for the set J. Moreover, $\operatorname{diam} \Delta_{w_0\dots w_{i-1}} = \lambda_{w_0} \cdots \lambda_{w_{i-1}}$ and

$$\operatorname{dist}(\Delta_{\underline{\omega}0}, \Delta_{\underline{\omega}1}) = (1 - \lambda_0 - \lambda_1)\lambda_{w_0} \cdots \lambda_{w_{i-1}} > 0 \tag{3.3}$$

where $\operatorname{dist}(x, y) = |x - y|$. Thus, J is resulting from a Moran construction. □

The basic sets of the first free generations for a Moran construction governed by the golden mean subshift are shown on Figure 3.2.

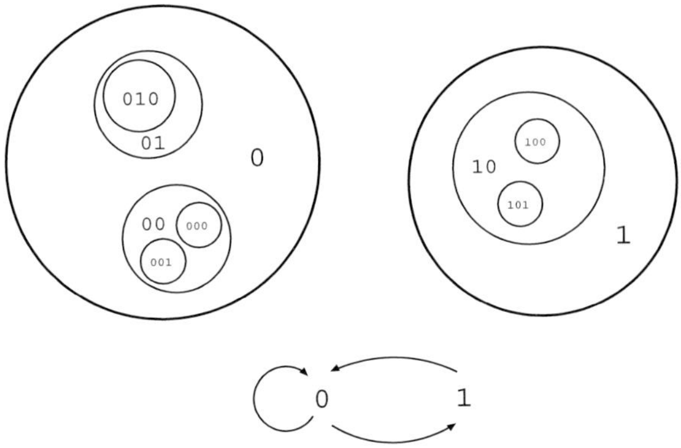

Figure 3.2. Moran construction governed by the golden mean subshift. At the length-three level, only the words 000, 001, 010, 100 and 101 are admissible.

3.1.1. Generalized Moran constructions

Of course, generally, invariant sets in dynamical systems are results of constructions much more sophisticated that just the Moran ones (see for instance [97]). In this subsection we restrict our attention to so called generalized Moran constructions.

DEFINITION 3.1. Consider $\Omega \subset \Omega_p$ an arbitrary subshift with positive topological entropy and let $\lambda : \Omega \to [\lambda_{\min}, \lambda_{\max}] \subset \mathbb{R}^+$ be an arbitrary Hölder continuous positive function, such that $0 < \lambda_{\min} < \lambda_{\max} < 1$. A geometric construction where Axiom (M_4) has been replaced by the assumption that there are positive constants $\underline{c}$ and $\bar{c}$ such that

$$\operatorname{diam} \Delta_{\omega_0 \ldots \omega_{n-1}} \geqslant \underline{c} \inf_{\omega \in [\underline{\omega}]} \prod_{j=0}^{|\underline{\omega}|-1} \lambda(\sigma^j \omega) \tag{3.4}$$

and

$$\operatorname{diam} \Delta_{\omega_0 \ldots \omega_{n-1}} \leqslant \bar{c} \sup_{\omega \in [\underline{\omega}]} \prod_{j=0}^{|\underline{\omega}|-1} \lambda(\sigma^j \omega) \tag{3.5}$$

is called a generalized Moran construction. In conditions (3.4) and (3.5) the cylinder set $[\omega_0 \ldots \omega_{n-1}]$ is denoted by $[\underline{\omega}]$.

Chapter 6 deals with generalized Moran constructions having $\lambda_{\max} = 1$.

LEMMA 3.1. *There exists a positive constant $\underline{d}$ such that*

$$\underline{d} \prod_{j=0}^{|\omega|-1} \lambda(\sigma^j \omega) \leqslant \left|\chi\left([\underline{\omega}]\right)\right|, \tag{3.6}$$

for every $\omega \in [\underline{\omega}]$.

We make use of the following sublemma in the proof of Lemma 3.1.

SUBLEMMA 3.1. *Fix $r > 0$. Consider a finite collection $\mathcal{D}$ of closed pairwise disjoint balls $B \subset \mathbb{R}^d$ with diameters $|B| \geqslant r$. Let $br := \max\{|B|\colon B \in \mathcal{D}\}$. If $\#\mathcal{D} \geqslant (2+b)^d$ there exist balls B_1, $B_2 \in \mathcal{D}$ such that* $\operatorname{dist}(B_1, B_2) \geqslant r$.

PROOF. Let $B_1 \in \mathcal{D}$ be a ball of diameter br. Let $A(br, br+2r)$ be an annulus in $\mathbb{R}^d$ around B_1 with inner diameter br and outer diameter $br + 2r$. See Figure 3.3

Denote by $\operatorname{Vol}(A(br, br+2r))$ and $\operatorname{Vol}(B(r))$ the volumes of the annulus and of a ball of diameter r, respectively. If we assume that we are in the case that

$$\begin{aligned} \frac{\operatorname{Vol}(A(br, br+2r))}{\operatorname{Vol}(B(r))} + 1 &= \frac{(br+2r)^d - (br)^d}{(r)^d} + 1 \\ &< (2+b)^d < \#\mathcal{D}. \end{aligned} \tag{3.7}$$

Then, for every positions of balls belonging to $\mathcal{D}$ within the annulus $A(br, br+2r)$ (even for "optimal" packing) there exists at least one ball $B_2 \in \mathcal{D}$ that is left outside $A(br, br+r)$. This proves the sublemma. □

PROOF OF LEMMA 3.1. We define, for an arbitrary admissible word $\underline{\omega} := (\omega_0 \dots \omega_{n-1})$, a radius $r > 0$ as follows. For fixed $m > 0$ let $r = r(m, \underline{\omega}) := \lambda_{\min}^m |\underline{B}(\underline{\omega})|$, where $\lambda_{\min} := \min_{\omega \in S}\{\lambda(\omega)\}$.

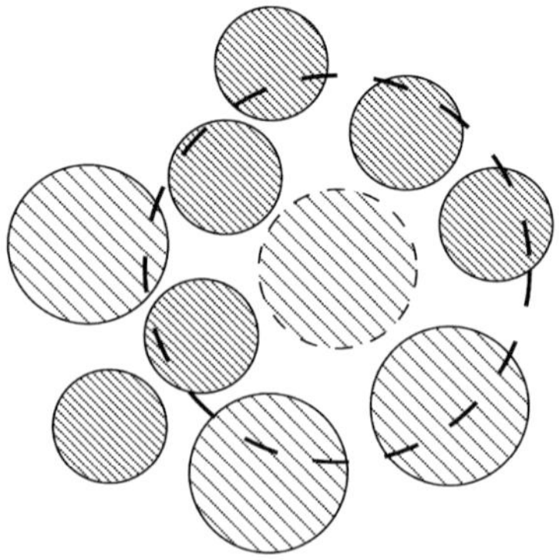

Figure 3.3. Balls of diameters r and br, with $1 < b < 2$.

We construct a finite collection of disjoint balls $\mathcal{D}_r(\underline{\omega})$ intersecting the set $\chi([\underline{\omega}])$ with controlled diameters. Let

$$\widetilde{\mathcal{D}}_r(\underline{\omega}) := \{\underline{B}(\underline{\omega}\,\underline{\omega}')\colon\ [\underline{\omega}\,\underline{\omega}'] \subset S\}$$

such that there exists an $a \in \{0, \ldots, p-1\}$ for which $\underline{\omega}\,\underline{\omega}'a$ is admissible and

$$|\underline{B}(\underline{\omega}\,\underline{\omega}')| \geqslant r > |\underline{B}(\underline{\omega}\,\underline{\omega}'a)|. \tag{3.8}$$

The requested collection of pair-wise disjoint balls $\mathcal{D}_r(\underline{\omega})$ is obtained by eliminating $\underline{B}(\underline{\omega}\,\underline{\omega}'') \in \widetilde{\mathcal{D}}_r(\underline{\omega})$ if there exists a ball $\underline{B}(\underline{\omega}\,\underline{\omega}') \in \widetilde{\mathcal{D}}_r(\underline{\omega})$ such that $\underline{\omega}''$ is a prefix of $\underline{\omega}'$.

For any ball $\underline{B}(\underline{\omega}\,\underline{\omega}') \in \mathcal{D}_r(\underline{\omega})$ the inequality (3.8) is satisfied. Moreover, $\underline{B}(\underline{\omega}\,\underline{\omega}') \cap \chi([\underline{\omega}]) \neq \emptyset$.

By construction of the set $\mathcal{D}_r(\underline{\omega})$, for every

$$\underline{B}(\underline{\omega}\,\underline{\omega}') = \underline{B}(\omega_0, \ldots, \omega_{n-1}, \omega'_0, \ldots, \omega'_{k-1}) \in \mathcal{D}_r(\underline{\omega})$$

we have

$$r \leqslant |\underline{B}(\underline{\omega}\,\underline{\omega}')| < \frac{r}{\lambda_{\min}}\,\frac{\bar{c}}{\underline{c}}. \tag{3.9}$$

Indeed, it exists an $a \in \{0, 1, \ldots, p-1\}$ such that $r > |\underline{B}(\underline{\omega}\,\underline{\omega}'a)|$. By condition (M2) in the Moran construction of the set F, for all $\omega \in [\underline{\omega}\,\underline{\omega}'a]$ we have

$$r > \underline{c}\,\lambda(\sigma^{n+k+1}\omega)\left(\prod_{j=1}^{n+k}\lambda(\sigma^j\omega)\right) \geqslant \lambda_{\min}\frac{\underline{c}}{\bar{c}}\,|\underline{B}(\underline{\omega}\,\underline{\omega}')|.$$

Since $k := |\underline{\omega}'| \leqslant m$ for all $\underline{B}(\underline{\omega}\,\underline{\omega}') \in \mathcal{D}_r(\underline{\omega})$ inequality (3.9) follows. As in Sublemma 3.1 we set $b := \bar{c}/(\underline{c}\,\lambda_{\min})$ and the first assumption of Sublemma 3.1 is satisfied for the set of balls $\mathcal{D}_r(\underline{\omega})$. The specification property and positive topological entropy imply that $\#\mathcal{D}_{r(m,\underline{\omega})}(\underline{\omega})$ grows exponentially as $m \to \infty$, independently of the choice of $\underline{\omega}$. So, we can choose m^* to be such that

$$\#\mathcal{D}_{r(m^*,\underline{\omega})}(\underline{\omega}) > \left(2 + \frac{\bar{c}}{\underline{c}\,\lambda_{\min}}\right)^d = (2+b)^d$$

for each admissible $\underline{\omega}$. Then according to Sublemma 3.1 for each $\underline{\omega}$ there exist balls $\underline{B}_1$ and $\underline{B}_2 \in \mathcal{D}_{r(m^*,\underline{\omega})}(\underline{\omega})$ for which $\operatorname{dist}(\underline{B}_1, \underline{B}_2) \geqslant r(m^*, \underline{\omega})$. Therefore, for all admissible $\underline{\omega}$ the following inequalities hold,

$$|\chi([\underline{\omega}])| \geqslant \operatorname{dist}(\underline{B}_1, \underline{B}_2) \geqslant \lambda_{\min}^{m^*}|\underline{B}(\underline{\omega})| \geqslant \lambda_{\min}^{m^*}\underline{c}\prod_{i=0}^{n-1}\lambda(\sigma^i\omega).$$

Thus the lemma is proved with $\underline{d} := \lambda_{\min}^{m^*}\,\underline{c}$. □

3.1.2. Invariant subsets of Markov maps

A left continuous function T defined on the interval $I = [x_0, x_p]$ is a Markov map when the following properties are satisfied.

(m$_1$) $T([x_0, x_p]) \subset [x_0, x_p]$.
(m$_2$) The map T is continuous at x_0.
(m$_3$) There exists a collection of intervals, called basic intervals, $\{I_i := [x_i, x_{i+1}]\}_{i=0}^{p-1}$, such that, for each i, the following conditions hold:
  (a) T is $C^{1+\varepsilon}$ on $\mathrm{Int}(I_i)$ and $\inf\{|T'(x)|: x \in \mathrm{Int}(I_i)\} > 1$.
  (b) For every $0 \leqslant j < p$ either $T(\mathrm{Int}(I_i)) \supset \mathrm{Int}(I_j)$ or $T(\mathrm{Int}(I_i)) \cap \mathrm{Int}(I_j) = \emptyset$.

When T is a Markov map, the system (I, T) is conjugated to a topological Markov chain (Ω_A, σ) with transition matrix A determined by condition (m$_3$) [104]. That is to say, there exists a continuous coding map $\chi : \Omega_A \to I$ such that $T \circ \chi = \chi \circ \sigma$. To specify the map χ, for each admissible word $\omega_0 \dots \omega_n$, $n \geqslant 0$, consider the intervals

$$I_{\omega_0 \dots \omega_n} := \bigcap_{k=0}^{n} \overline{T^{-k}(I_{\omega_k})} = \overline{\bigcap_{k=0}^{n} T^{-k}(I_{\omega_k})},$$

where the second equality follows from property (m$_3$). The coding map is expressed as follows

$$\chi(\omega) = \bigcap_{k=0}^{\infty} I_{\omega_0 \dots \omega_k}.$$

For every $n \geqslant 0$, every basic interval I_{ω_0}, can be written as the following union

$$I_{\omega_0} = \bigcup_{[\omega_0 \dots \omega_n] \subset \Omega_A} I_{\omega_0 \dots \omega_n}. \tag{3.10}$$

The intervals $I_{\omega_0 \dots \omega_n}$ in the previous union intersect each other at most at their boundaries. Moreover, let $s(i) = 1$ if $T'(x) > 0$ for $x \in \mathrm{Int}(I_i)$ and let $s(i) = -1$ otherwise. Then (Ω_A, s) is totally ordered and reflects the order in I, i.e., $\mathrm{Int}(I_{\omega_0 \dots \omega_n}) < \mathrm{Int}(I_{\varpi_0 \dots \varpi_n})$ iff $\omega_0 \dots \omega_n < \varpi_0 \dots \varpi_n$.

Let T be a Markov map for which the corresponding ordered topological Markov chain (Ω, σ, s) is mixing. A T-invariant Cantor set, Λ_n, is obtained by a generalized Moran construction. Indeed, for each $\omega_0 \in \{0, \dots, p-1\}$ and positive n let $I_{\omega_0 \hat{\omega}_1 \dots \hat{\omega}_n}$ be the leftmost interval in the union (3.10) and let B_n denote the set of words corresponding to the leftmost intervals:

$$B_n := \{\omega_0 \hat{\omega}_1 \dots \hat{\omega}_n,\ \omega_0 = 0, \dots, p-1\}.$$

The set of admissible sequences not containing any word in B_n,

$$\widehat{\Omega}_n = \{\omega \in \Omega\colon\ \omega_k \dots \omega_{k+n} \notin B_n,\ k \in \mathbb{Z}^+\}$$

defines a symbolic system $(\widehat{\Omega}_n, \sigma)$ which is a topological Markov chain on words of length $n+1$. This system may not be mixing but it contains a mixing topological Markov chain, (Ω_n, σ), if n is sufficiently large. This topological Markov chain is specified by the set C_n that collects all words in B_n^c that are pair-wise Ω_n-connected. Such a subshift was constructed in Section 2.2.

For (Ω_n, σ) and $k > n$ consider basic intervals $\Delta_{\omega_0\omega_1\dots\omega_{k-1}} \subset [x_0, x_p]$ as follows. First, for each $\omega_0\omega_1 \dots \omega_n \in C_n$ let $I_{\omega_0\omega_1\dots\omega_n}$ be declared to be the basic intervals $\Delta_{\omega_0\omega_1\dots\omega_n}$ and let $\Delta = \bigcup_{\underline{\omega} \in C_n} \Delta_{\underline{\omega}}$. Then, for given $i = 0, \dots, p-1$ and $\underline{\omega} \in C_n$ let $K^i_{\underline{\omega}} = \Delta_{\underline{\omega}}$ if

$$T\Bigl(\bigcup_{i\varpi_1\dots\varpi_n \in C_n} \Delta_{i\varpi_1\dots\varpi_n}\Bigr) \supset \Delta_{\underline{\omega}}$$

and $K^i_{\underline{\omega}} = \emptyset$ otherwise. Let $\Delta^i = \bigcup_{\underline{\omega} \in C_n} K^i_{\underline{\omega}}$ and let $w_i : \Delta^i \to \Delta \cap [x_i, x_{i+1}]$ be such that $T(w_i(x)) = x$. Then, for $k > n$ and every word $\omega_0 \dots \omega_{n+1} \dots \omega_{k-1}$ that is admissible in Ω_n, the non-empty set

$$\Delta_{\omega_0\dots\omega_n\omega_{n+1}\dots\omega_{k-1}} = w_{k-1} \circ \dots \circ w_{n+1}(\Delta_{\omega_0\dots\omega_n})$$

is a basic interval.

The map $\lambda := \frac{1}{|T'|} \circ \chi$ is Hölder continuous in Ω_A and for any $\omega \in \Omega_A$ and any $n \in \mathbb{Z}^+$ the inequalities

$$\rho^{-1} \prod_{i=0}^{n} \lambda(\sigma^i \omega) \leqslant \operatorname{diam} \Delta_{\omega_0\dots\omega_n} \leqslant \rho \prod_{i=0}^{n} \lambda(\sigma^i \omega), \tag{3.11}$$

hold for some $\rho \geqslant 1$. Since $\Omega_n \subset \Omega_A$, we are dealing with a generalized Moran construction.

The dynamical systems $(\Lambda_n, T|\Lambda_n)$, on the limit set Λ_n, and the mixing subshift (Ω_n, σ) are conjugated. The map $\chi|\Omega_n$ provides the conjugacy. Moreover, since Ω_n is compact, totally disconnected and does not contain isolated points, so does the T-invariant subset $\Lambda_n \subset [x_0, x_p]$.

EXAMPLE 3.2 *(The tent map)*. The previous construction is exemplified for the tent map, $T : [0, 1] \to [0, 1]$, defined as $T(x) = \min\{2x, 2(1-x)\}$. It is a Markov map with basic partition $I_0 = [x_0, x_1] = [0, 1/2]$ and $I_1 = [x_1, x_2] = [1/2, 1]$. Figure 3.4 shows the refinement of this partition as specified by words of length $n = 3$, namely $I_{w_0w_1w_2}$. At the length $n = 3$, the leftmost subintervals in I_0 and I_1 are I_{000} and I_{110}, respectively. In the symbolic description, removing the intervals I_{000} and I_{110} corresponds to forbid the words in $B_2 = \{000,\ 110\}$. The subshift

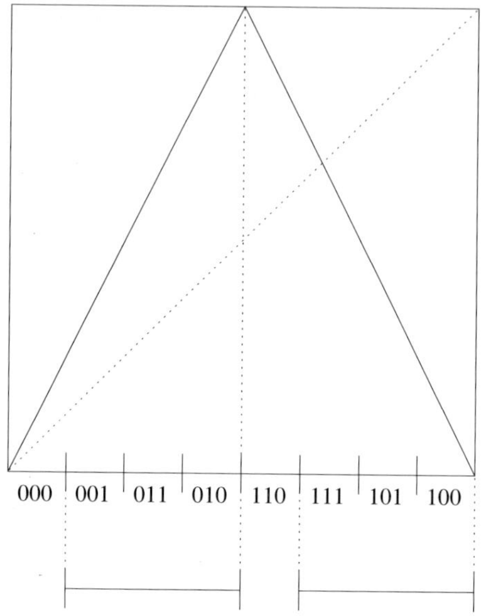

Figure 3.4.

$(\widetilde{\Omega}_3, \sigma)$ thus defined is specified by the graph shown in Figure 3.5. This subshift is not mixing. However, the subgraph spanned by the vertices enclosed in a box (corresponding to words in C_3) specifies a mixing subshift (Ω_3, σ). Each word in C_3 corresponds to the interval in Figure 3.4 that is labeled by the same word. The Cantor set Λ_3 resulting of the generalized Moran construction is the image of the subshift Ω_3 under the semiconjugacy $\chi|\Omega_3$.

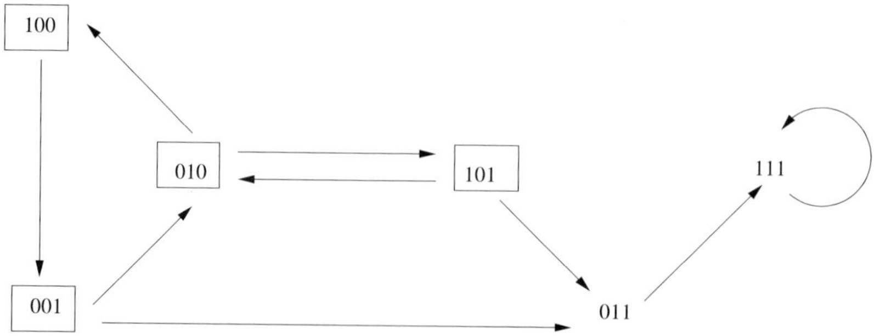

Figure 3.5.

3.2. Topological pressure and Hausdorff dimension

One of the main characteristics of invariant sets is the Hausdorff dimension. Let us remind its definition.

3.2.1. *Hausdorff and box dimensions*

Let X be a metric space with a distance $d(x, y)$ between points x and $y \in X$. For any subset $A \subset X$ let $\{U_i\}$ be a finite or countable collection of open sets of diameter less than ε such that $\bigcup U_i \supset A$; here

$$\operatorname{diam} U_i := \sup\{d(x, y)\colon\ x, y \in U_i\}.$$

For any $\alpha > 0$ we introduce

$$m(\alpha, \varepsilon, A) = \inf_{\{U_i\}} \sum_i (\operatorname{diam} U_i)^\alpha, \tag{3.12}$$

where infimum is taken over all covers $\{U_i\}$ with diameter less than ε, and

$$m(\alpha, A) = \lim_{\varepsilon \to 0} m(\alpha, \varepsilon, A), \tag{3.13}$$

the α-dimensional Hausdorff measure (the limit exists because of monotonicity of $m(\alpha, \varepsilon, A)$ as a function of ε). It is simple to see that $m(\beta, \varepsilon, A) \leqslant \varepsilon^{\beta-\alpha} m(\alpha, \varepsilon, A)$, which implies that there exists a unique critical value α_c of α such that $m(\alpha, A) = 0$ if $\alpha > \alpha_c$ and $m(\alpha, A) = \infty$ if $\alpha < \alpha_c$. The quantity $\alpha_c =: \dim_H A$ is called the Hausdorff dimension.

EXAMPLE 3.3. For the set J constructed in Example 3.1 we have that $\dim_H J = s_0$, where s_0 is the root of the equation

$$\lambda_0^s + \lambda_1^s = 1. \tag{3.14}$$

To make (3.14) evident, consider the cover of J by basic sets of the n-th generation. Then the sum $\sum_i (\operatorname{diam} U_i)^\alpha$ in (3.12), up to a constant, becomes

$$\sum_{\omega_0, \dots, \omega_{n-1}} \prod_{k=0}^{n-1} \lambda_{\omega_k}^\alpha = \left(\lambda_0^\alpha + \lambda_1^\alpha\right)^n. \tag{3.15}$$

If $\alpha > s_0$, then (3.15) goes to zero as $n \to \infty$, that shows us that $\dim_H J \leqslant s_0$. To get the opposite inequality, people use the technique of so-called Moran covers [97], see below.

Moran proved in [87] that for geometric constructions satisfying conditions (M_1)–(M_4) the Hausdorff dimension is $\dim_H F = s_0$, where s_0 is the root of the

(Moran) equation

$$\sum_{i=0}^{p-1} \lambda_i^s = 1. \tag{3.16}$$

Similar formulas could be obtained in the case when not all words are admissible, in particular, in the case of subshifts. In these cases Hausdorff dimensions of invariant sets can be expressed in terms of topological pressure.

An important notion often people use is the box dimension. In the sum (3.12) one may consider open sets of diameter equals ε, i.e.,

$$m(\alpha, \varepsilon, A) = N_\varepsilon \varepsilon^\alpha$$

where N_ε is the number of such sets. The limit of $m(\alpha, \varepsilon, A)$ as ε goes to 0 may not exist, therefore one defines

$$\overline{m}(\alpha, A) = \limsup_{\varepsilon \to 0} m(\alpha, \varepsilon, A)$$

and

$$\underline{m}(\alpha, A) = \liminf_{\varepsilon \to 0} m(\alpha, \varepsilon, A).$$

The upper box dimension

$$\overline{\dim}_B A = \sup\{\alpha\colon \overline{m}(\alpha, A) = \infty\}$$

is called in literature the fractal dimension of set A. The lower box dimension is defined as

$$\underline{\dim}_B A = \sup\{\alpha\colon \underline{m}(\alpha, A) = \infty\}.$$

If $\overline{\dim}_B A = \underline{\dim}_B A =: b$ then b is called the box dimension. It is simple to see that the box dimension b can be defined as

$$B = \lim_{\varepsilon \to 0} \frac{\log N_\varepsilon}{-\log \varepsilon}$$

where N_ε is the number of balls of radius ε needed to cover the set A. The upper and lower box dimensions are examples of upper and lower Carathéodory capacities defines in Section 4.3.

Conditions (M_1)–(M_4) provide a more general geometric scenario than the one presented in [100]. The conditions (M_1)–(M_3) can be seen as a particular case of the subadditive formalism developed in [15]. In all these cases the Hausdorff and box dimensions of the set F coincide and are equal to the root of an equation of the Bowen type.

However, we will not follow the way described in [97,15]. For our goal it is more appropriate to impose conditions (3.4) and (3.5) and describe the results in a more clear form.

3.2.2. Bowen's equation

Let us show now how the topological pressure is related to the Hausdorff dimension. Assume first that a set F is modeled by a Moran construction and the corresponding subshift is a topological Markov chain (σ, Ω_A). Choose a cover of F by basic sets of the n-th generation. Then,

$$\begin{aligned}\sum_{\omega_0\ldots\omega_{n-1}} (\operatorname{diam} \Delta_{\omega_0\ldots\omega_{n-1}})^\alpha &= \sum_{\omega_0\ldots\omega_{n-1}} \prod_{k=0}^{n-1} \lambda_{\omega_k}^\alpha \\ &= \sum_{\omega_0\ldots\omega_{n-1}} \exp\left(\alpha \sum_{j=0}^{n-1} \varphi(\omega_j)\right) \\ &= Z_n(\alpha\varphi, \Omega_A)\end{aligned} \tag{3.17}$$

where $\varphi(\omega_0, \omega_1, \ldots) = \log \lambda_{\omega_0}$. We know that

$$Z_n(\alpha\varphi, \Omega_A) \approx \exp\bigl(n P_{\Omega_A}(\alpha\varphi)\bigr).$$

Hence, $Z_n(\alpha\varphi, \Omega_A) \gg 1$ if $P_{\Omega_A}(\alpha\varphi) > 0$ and $Z_n(\alpha\varphi, \Omega_A) \ll 1$ if $P_{\Omega_A}(\alpha\varphi) < 0$. It follows that if α_0 is the root of the (Bowen's) equation

$$P_{\Omega_A}(\alpha\varphi) = 0 \tag{3.18}$$

then $\dim_H F \leqslant \alpha_0$. The opposite inequality can be proven by using the technique of Moran covers and the dimension-like definition of topological pressure [97], see below.

The Bowen's equation holds not only for such a simple potential function φ as the one considered in (3.17), but also in the case of generalized Moran constructions. To prove it we need the notion of the Moran covers.

3.2.3. Moran covers

We describe them in slightly different form than in [97]. Given an open ball $B \subset \mathbb{R}^m$, a basic set $\Delta_{\omega_0\ldots\omega_{n-1}}$ is called B-related if $\Delta_{\omega_0\ldots\omega_{n-1}} \cap B \neq \emptyset$, $\operatorname{diam} \Delta_{\omega_0\ldots\omega_{n-2}} \geqslant \operatorname{diam} B$, but $\operatorname{diam} \Delta_{\omega_0\ldots\omega_{n-1}} < \operatorname{diam} B$. Let $R(B)$ be the collection of all B-related basic sets. It is known that if $\operatorname{diam}|B| \ll 1$ then $\#R(B) \leqslant M$ where M is constant depending only on m. Therefore,

$$(\operatorname{diam} B)^\alpha \geqslant \frac{1}{M} \sum_{\Delta^j \in R(B)} (\operatorname{diam} \Delta^j)^\alpha \tag{3.19}$$

for any non-negative α. We consider now an arbitrary finite cover of F by balls B_i of diameters $\varepsilon_i < \varepsilon$, $i = 0, \ldots, N-1$. Then, collection $R(B_i)$, $i = 0, \ldots, N-1$, form a cover, say $\mathcal{C}$, of F which is called the Moran cover. It is a cover by cylinders but may be of different lengths. Because of the inequality (3.19), we have

$$\sum_{i=0}^{N-1} \varepsilon_i^{\alpha} \geqslant \frac{1}{M} \sum_{i=0}^{N-1} \sum_{\Delta_{\omega_0 \dots \omega_k} \in R(B_i)} (\operatorname{diam} \Delta_{\omega_0 \dots \omega_k})^{\alpha}, \tag{3.20}$$

where the second sum is taken over all B_i-related basic sets. Given $\varepsilon > 0$, there is $n = n(\varepsilon)$ such that for any B_i-related basic set $\Delta_{\omega_0 \dots \omega_k}$ we have $k > n(\varepsilon)$. Moreover, $n(\varepsilon) \to \infty$ as $\varepsilon \to 0$. By using (3.17), we obtain that the right hand side of (3.20) is bounded from below by

$$\frac{1}{M} \sum_{\omega_0 \dots \omega_{n-1}} \prod_{k=0}^{n-1} \lambda_{\omega_k}^{\alpha} = \frac{1}{M} \sum_{\omega_0 \dots \omega_{n-1}} \exp\left(\alpha \sum_{k=0}^{n-1} \log \lambda_{\omega_k} \right) \tag{3.21}$$

where the sum is taken over all words $\omega_0 \dots \omega_{n-1}$ corresponding to B_i-related basic sets, for all i. We use (3.20) and (3.21) to prove the main result of this subsection.

THEOREM 3.1. *For a generalized Moran construction* $\dim_H F = \alpha_c$, *where* α_c *is the root of the Bowen's equation*

$$P_{\Omega}(\alpha\varphi) = 0,$$

with $\varphi(\omega) = \log \lambda(\omega)$.

Let us rewrite the statistical sum (2.16) for this particular case:

$$\mathcal{Z}(\beta, \alpha \log \lambda_{\omega_k}, \mathcal{C}, \Omega) = \sum_{\Delta_{\underline{\omega}} \in \mathcal{C}} \exp\left(-\beta |\underline{\omega}| + \alpha \sum_{k=0}^{|\underline{\omega}|-1} \log \lambda_{\omega_k} \right). \tag{3.22}$$

Assume that $\alpha < \beta_c = P_{\Omega}(\alpha \log \lambda_0)$, then, for any $K > 0$, there is $n_0 = n_0(K)$ such that $\mathcal{Z}(\beta_c, \alpha \log \lambda_{\omega_k}, \mathcal{C}, \Omega) > K$ provided that $n(\varepsilon) > n_0$. It follows that

$$\sum_{\Delta_{\underline{\omega}} \in \mathcal{C}} \exp\left(\alpha \sum_{k=0}^{|\underline{\omega}|-1} \log \lambda_{\omega_k} \right) \geqslant K e^{\beta_c n(\varepsilon)} \tag{3.23}$$

i.e., the inequalities (3.20) and (3.21) imply that

$$\sum_{i=0}^{N-1} \varepsilon_i^{\alpha} \geqslant \frac{K}{M} e^{\beta_c n(\varepsilon)}. \tag{3.24}$$

This implies that $\dim_H F \geqslant P_{\Omega}(\alpha \log \lambda_0)$. The opposite inequality has been already obtained in Section 3.2.2. Thus, we proved that $\dim_H F = \alpha_0$, the root of the Bowen equation (3.18).

EXAMPLE 3.4. Let us come back to Example 3.1. In this case

$$A = \begin{pmatrix} 1 & 1 \\ 1 & 1 \end{pmatrix},$$

$0 < \lambda_{0,1} < 1$ are rates of contraction,

$$\psi(\omega_0, \omega_1, \dots) = \alpha \log \lambda_{\omega_0} = \alpha\varphi(\omega_0, \omega_1, \dots).$$

Thus, $\rho_i = \lambda_i^\alpha$, $i = 0, 1$,

$$B = \begin{pmatrix} \lambda_0^\alpha & \lambda_1^\alpha \\ \lambda_0^\alpha & \lambda_1^\alpha \end{pmatrix}$$

and $P_{\Omega_2}(\alpha\varphi) = \log(\lambda_0^\alpha + \lambda_1^\alpha)$. The Bowen's equation (3.18) becomes the Moran's equation (3.14).

EXAMPLE 3.5. Consider now the "golden mean" topological Markov chain with the transition matrix

$$A = \begin{pmatrix} 1 & 1 \\ 1 & 0 \end{pmatrix}.$$

Assume that $0 < \lambda_{0,1} < 1$ are rates of contraction. Here again $\rho_i = \lambda_i^\alpha$ but the matrix B has the form

$$B = \begin{pmatrix} \lambda_0^\alpha & \lambda_1^\alpha \\ \lambda_0^\alpha & 0 \end{pmatrix}.$$

The characteristic equation of the matrix B is $\mu^2 - \mu\lambda_0^\alpha - (\lambda_0\lambda_1)^\alpha = 0$ and spectral radius is

$$r = \frac{1}{2}\left(\lambda_0^\alpha + \sqrt{\lambda_0^{2\alpha} + 4(\lambda_0\lambda_1)^\alpha}\right).$$

Thus, the Hausdorff dimension of the corresponding set F is the root of the Bowen's equation

$$\log \frac{1}{2}\left(\lambda_0^\alpha + \sqrt{\lambda_0^{2\alpha} + 4(\lambda_0\lambda_1)^\alpha}\right) = 0.$$

If $\lambda_0 = \lambda_1 = \lambda$ then the equation becomes $\alpha \log \lambda + \log((1 + \sqrt{5})/2)$, i.e.,

$$\dim_H F = \alpha_0 = \frac{\log((1 + \sqrt{5})/2)}{-\log \lambda}.$$

If you take into account that $\log((1 + \sqrt{5})/2) = h_{\text{top}}$, the topological entropy of the topological Markov chain (σ, Ω_A), then we obtain the relation ([57])

$$\dim_H F = \frac{h_{\text{top}}}{-\log \lambda}.$$

3.3. Strong Moran construction

For calculations of dimensions for Poincaré recurrences below, we need to define Moran constructions satisfying some additional conditions. A subclass of Moran constructions, satisfying (M_1)–(M_3), is defined by adding the following "gap condition". There is a (gap) constant $G > 0$ such that for all admissible words $(\omega_0, \dots, \omega_{i-1}, \omega_i)$ and $(\omega_0, \dots, \omega_{i-1}, \omega_i')$ one has

$$\operatorname{dist}(\Delta_{\omega_0\dots\omega_{i-1}\omega_i}, \Delta_{\omega_0\dots\omega_{i-1}\omega_i'}) \geqslant G \operatorname{diam} \Delta_{\omega_0\dots\omega_{i-1}}, \tag{3.25}$$

$$\operatorname{dist}(\Delta_{\omega_0}, \Delta_{\omega_0'}) \geqslant G, \tag{3.26}$$

if $\omega_i \neq \omega_i'$ in (3.25) and $\omega_0 \neq \omega_0'$ in (3.26). Moran constructions satisfying condition (3.25) and (3.26) are said to be strong Moran constructions and the corresponding fractal sets F are said to satisfy a gap condition.

3.4. Controlled packing of cylinders

The following notion is very close to the notion of B-related cylinders.

Given an open ball $B \in \mathbb{R}^d$, a cylinder $[\omega_0, \omega_1, \dots, \omega_{n-1}]$ is called B-maximal if and only if $\chi([\omega_0, \omega_1, \dots, \omega_{n-1}]) \subset F \cap B$ and $\chi([\omega_0, \omega_1, \dots, \omega_{n-2}]) \not\subset F \cap B$. The set of all B-maximal cylinders is denoted by $\operatorname{CMax}(B)$. Let $\mathcal{C}$ be a cover of F by sets out of $\mathcal{B}_\varepsilon$. The collection of all B-maximal cylinders in $\mathcal{C}$,

$$\operatorname{CMax}(\mathcal{C}) := \bigcup_{B \in \mathcal{C}} \operatorname{CMax}(B),$$

is a cover of S by cylinders. We say that $F \subset \mathbb{R}^d$ has controlled packing of cylinders if there exist positive constants C_0 and a, independent of ε and $\mathcal{C}$, such that for every open ball $B \in \mathcal{C}$, every $0 < \rho < 1$ and every positive integer N one has

$$\#\{[\underline{\omega}] \in \operatorname{CMax}(B)\colon \left|\chi([\underline{\omega}])\right| \in (\rho^{N+1}, \rho^N]\} \leqslant C_0 N^a. \tag{3.27}$$

A fractal set F resulting from a Moran construction satisfying (3.27) is said to have the controlled-packing property.

We show next that for dimension $d = 1$ fractal sets always satisfy the controlled-packing condition (3.27). Consider a (non-strong) Moran construction where the basic Δ-sets belong to an interval.

LEMMA 3.2. *Let F be the limit set of a one-dimensional Moran construction. Then F has controlled packing of cylinders: for every $0 < \rho < 1$ there exists a non-negative constant C_0 such that*

$$\#\{[\underline{\omega}] \in \operatorname{CMax}(B)\colon \left|\chi([\underline{\omega}])\right| \in (\rho^{N+1}, \rho^N]\} \leqslant C_0 N. \tag{3.28}$$

PROOF. In the one-dimensional case for any admissible word $(\omega_0, \ldots, \omega_{n-1})$, $n \in \mathbb{N}$, we have that

$$\overline{B}(\omega_0, \ldots, \omega_{n-1}) = \underline{B}(\omega_0, \ldots, \omega_{n-1}) = \Delta(\omega_0, \ldots, \omega_{n-1})$$

is an interval and $\bar{c} = \underline{c} = 1$. Thus, the sets $\chi([\underline{\omega}]_n)$ are ordered along the line and therefore every interval B containing $2p - 1$ sets of the form $\chi([\underline{\omega}]_n)$ contains also at least one set of the form $\chi([\underline{\omega}]_{n-1})$. This implies that

$$\#\bigl\{[\underline{\omega}] \in \mathrm{CMax}(B)\colon |\underline{\omega}| = n\bigr\} \leqslant 2p - 2,$$

for any $n > 0$. Using this result and the inequality (3.6), we obtain the following estimate

$$\begin{aligned} &\#\bigl\{[\underline{\omega}] \in \mathrm{CMax}(B)\colon \bigl|\chi\bigl([\underline{\omega}]\bigr)\bigr| \in \bigl(\rho^{N+1}, \rho^{N}\bigr]\bigr\} \\ &\quad \leqslant (2p-2)\,\#\bigl\{n\colon \bigl(\rho^{N+1}, \rho^{N}\bigr] \cap \bigl[\underline{d}\lambda^n_{\min}, \lambda^n_{\max}\bigr] \neq \emptyset\bigr\}, \end{aligned} \tag{3.29}$$

where $\lambda_{\max} := \max_{\omega \in S}\{\lambda(\omega)\}$. Direct calculations show that if

$$\bigl(\rho^{N+1}, \rho^{N}\bigr] \cap \bigl[\underline{d}\lambda^n_{\min}, \lambda^n_{\max}\bigr] \neq \emptyset$$

then

$$\frac{N \log \rho - \log \underline{d}}{\log \lambda_{\min}} =: n_{\min} \leqslant n < n_{\max} := \frac{N \log \rho}{\log \lambda_{\max}} + \frac{\log \rho}{\log \lambda_{\max}}$$

and

$$\begin{aligned} &\#\bigl\{n\colon \bigl(\rho^{N+1}, \rho^{N}\bigr] \cap \bigl[\underline{d}\lambda^n_{\min}, \lambda^n_{\max}\bigr] \neq \emptyset\bigr\} \\ &\quad \leqslant N\left(\frac{\log \rho}{\log \lambda_{\max}} - \frac{\log \rho}{\log \lambda_{\min}}\right) + \frac{\log \rho}{\log \lambda_{\max}} + \frac{\log \underline{d}}{\log \lambda_{\min}}. \end{aligned} \tag{3.30}$$

Thus, we obtain (3.28) with $C_0 \geqslant \overline{C}_0 + B_0$ where

$$\overline{C}_0 = (2p-2)\left(\frac{\log \rho}{\log \lambda_{\max}} - \frac{\log \rho}{\log \lambda_{\min}}\right) \tag{3.31}$$

and

$$B_0 = (2p-2)\left(\frac{\log \rho}{\log \lambda_{\max}} + \frac{\log \underline{d}}{\log \lambda_{\min}}\right). \tag{3.32}$$

□

3.5. Sticky sets

An area preserving map f of the plane possessing an infinite hierarchy of islands-around-islands structure has invariant sets of zero Lebesgue measure on which it behaves similarly to multipermutative systems [1,11].[1] It was shown in Sec-

[1] Poincaré recurrences in the phase space of area preserving maps are analyzed in detail in Chapter 15.

tion 2.3 that every minimal multipermutative system (Ω, T) is topologically conjugate to the p-adic adding machine. The map T is not chaotic and its topological entropy is zero. A set F on which f is topologically conjugate to T, nevertheless, may appear as a result of a Moran type geometric construction.

Sticky sets are the sets of all limiting points of infinite hierarchy of islands-around-islands. A closed topological disk $\mathcal{I}$ is said to be an island of stability if $f^n(\mathcal{I}) = \mathcal{I}$ for some integer n. We now give a definition of infinite hierarchy of islands-around-islands structure (sticky riddle) for the general case when not all words $\underline{\omega} = (\omega_0, \dots, \omega_{n-1})$ might be admissible.

A collection $\mathcal{R}$ of islands $\{\mathcal{I}_{\underline{\omega}}\colon \underline{\omega} \text{ is } \Omega\text{-admissible}\}$ is said to be a sticky riddle if the sets $\mathcal{I}_{\underline{\omega}}$ are pairwise disjoint, are contained in a compact set, and

(i) for any island $\mathcal{I}_{\underline{\omega}} \in \mathcal{R}$ there is an island $\mathcal{I}_{\underline{\varpi}} \in \mathcal{R}$, $|\underline{\omega}| = |\underline{\varpi}|$, such that $f(\mathcal{I}_{\underline{\omega}}) = \mathcal{I}_{\underline{\varpi}}$;
(ii) if $f(\mathcal{I}_{\underline{\omega}}) = \mathcal{I}_{\underline{\varpi}}$ then for any admissible word $\underline{\omega}k$ there is $s \in \{0, 1, \dots, q-1\}$ such that $f(\mathcal{I}_{\underline{\omega}k}) = \mathcal{I}_{\underline{\varpi}s}$;
(iii) diam $(\mathcal{I}_{\underline{\omega}}) \to 0$ as $|\underline{\omega}| \to \infty$;
(iv) for any $\omega = (\omega_0, \omega_1, \dots) \in \Omega$, if $x_n \in \mathcal{I}_{\omega_0,\dots,\omega_{n-1}}$, $n > 0$, then $\lim_{n\to\infty} x_n$ exists;
(v) if $x_n \in \mathcal{I}_{\underline{\omega}}$, $y_n \in \mathcal{I}_{\underline{\varpi}}$, $|\underline{\omega}| = |\underline{\varpi}| = n$, $n > 0$, and $\underline{\omega} \neq \underline{\varpi}$ at least for one value of n then $\lim_{n\to\infty} x_n \neq \lim_{n\to\infty} y_n$.

The following axioms reflect our understanding of an infinite islands-around-islands hierarchy:

(i) An island of the n-th generation is mapped into an island of the same generation.
(ii) If an island $\mathcal{I}_{\underline{\omega}k}$ lies in the vicinity of the island $\mathcal{I}_{\underline{\omega}}$ then its image $\mathcal{I}_{\underline{\varpi}s}$ lies in a vicinity of $\mathcal{I}_{\underline{\varpi}}$.
(iii) To be packed into a compact set, the islands of the nth generation should be small if $n \gg 1$.
(iv) there should be only one point of accumulation of islands $\mathcal{I}_{\omega_0,\dots,\omega_{n-1}}$ for any fixed $\omega = (\omega_0, \dots, \omega_{n-1}, \dots)$;
(v) for different points $\omega = (\omega_0, \omega_1, \dots)$, $\omega' = (\omega'_0, \omega'_1, \dots)$ in Ω the corresponding points of accumulation of islands should be different.

Let $\mathcal{R}$ be a sticky riddle. For any $\omega = (\omega_0, \omega_1, \dots) \in \Omega$ and any sequence $x_n \in \mathcal{I}_{\omega_0,\dots,\omega_{n-1}}$, define $x = x(\omega) := \lim_{n\to\infty} x_n$. The set $\Lambda = \{x(\omega) : \omega \in \Omega\}$ is said to be a sticky set. It is well defined thanks to Axioms (iii)–(v).

The following theorem has been proven in [1].

THEOREM 3.2. *The system $f|\Lambda$ is topologically conjugated to a multipermutative system.*

3.5.1. Geometric constructions of sticky sets

Some numerical observation [21] show that sometimes every island of stability $\mathcal{I}_{\underline{\omega}}$, together with all its satellites $\mathcal{I}_{\underline{\omega\varpi}}$, belongs to a basic set $\Delta_{\underline{\omega}}$ of a geometric construction. So, the set Λ can be resulted from this construction. Axiomatically, the conditions for that can be expressed as follows.

(1) There exists a collection of sets $\{\Delta_{\underline{\omega}}\colon \underline{\omega}$ is admissible$\}$ that are closed, and for each admissible word $\underline{\omega}$, $\mathcal{I}_{\underline{\omega\varpi}} \subset \Delta_{\underline{\omega}}$ for every admissible word $\underline{\omega\varpi}$.
(2) $\mathcal{I}_{\underline{\omega}} \cap \Delta_{\underline{\omega\varpi}} = \emptyset$ for every admissible $\underline{\omega}$ and $\underline{\omega\varpi}$.
(3) $\Delta_{\underline{\omega j}} \subset \Delta_{\underline{\omega}}$, for every admissible words $\underline{\omega}$ and $\underline{\omega j}$.
(4) $\operatorname{diam} \Delta_{\omega_0\ldots\omega_{n-1}} \to 0$ as $n \to \infty$.
(5) Separation axiom. $\Delta_{\underline{\omega}} \cap \Delta_{\underline{\varpi}} \cap F = \emptyset$ if $\underline{\omega} \neq \underline{\varpi}$, $|\underline{\omega}| = |\underline{\varpi}|$, where

$$F = \bigcap_{n=1}^{\infty} \bigcup_{\substack{\omega_0\ldots\omega_{n-1} \\ \text{is admissible}}} \Delta_{\omega_0\ldots\omega_{n-1}}$$

Thus, if these axioms are satisfied, then $\Lambda = F$. To calculate the Hausdorff dimension of the set F we further assume conditions (3.4) and (3.5). Then we have the following.

PROPOSITION 3.1. $\dim_H F = \underline{\dim}_B F = \overline{\dim}_B F = \alpha_c$, *where* α_c *is solution of the Bowen equation* $P_\Omega(\alpha\varphi) = 0$ *with* $\varphi(\omega) = \ln \lambda_{\omega_0}$.

If Ω is the full shift on p symbols, then α_c is the root of Moran's equation

$$\lambda_0^{\alpha_c} + \cdots + \lambda_{n-1}^{\alpha_c} = 1.$$

Let us emphasize that an invariant set with nonchaotic dynamics is resulted from a geometric construction, modeled by a subshift (Ω, σ) *with positive topological entropy*. In other words, we have a "contradiction" between temporal and spatial behavior of a system. To describe such a situation, we need characteristics which could take into account both temporal and spatial behavior. Such characteristics are introduced the following chapters. Space-time behavior of sticky sets is discussed in Chapter 15.

The examples of sticky sets and the construction for the Feigenbaum attractor (see, for instance, [116]) show us that, in general, we should apply a wider notion than the Hausdorff dimension to describe simultaneously behavior of orbits on invariant sets and their geometric origination. The generalized Carathéodory construction allows us to do it.

Chapter 4

The Spectrum of Dimensions for Poincaré Recurrences

To study spatio-temporal behavior of orbits one needs to take into account not only geometrical properties of an invariant set X but also some temporal characteristics of orbits. A way to "marry" these two features is based on the Carathéodory–Hausdorff approach [40,64,97]. For our exposition it is convenient to represent definitions and main examples in the form of tables. The main object is the quadruple $(\mathcal{F}, \psi, \xi, \eta)$, said to be a Carathéodory structure for X, where $\mathcal{F}$, ψ, ξ and η are defined in Table 4.1.

4.1. Generalized Carathéodory construction

Let X denote a set and let $A \subset X$. Let us consider a finite or countable cover $G = \{u_i\}$ of A by elements of $\mathcal{F}$, with $\psi(u_i) \leqslant \varepsilon$ (see Table 4.1). Then, introduce the sum

$$M_\xi(\alpha, \varepsilon, G, A) = \sum_i \xi(u_i)\eta(u_i)^\alpha$$

and consider its infimum

$$M_\xi(\alpha, \varepsilon, A) = \inf_G \sum_i \xi(u_i)\eta(u_i)^\alpha, \tag{4.1}$$

over all covers $G \subset \mathcal{F}$ of A with $\psi(U_i) \leqslant \varepsilon$. The quantity $M_\xi(\alpha, \varepsilon, A)$ is a monotone function in ε; therefore, there exists the limit

$$m_\xi(\alpha, A) = \lim_{\varepsilon\to 0} M_\xi(\alpha, \varepsilon, A).$$

It was shown in [97] that there exists a critical value $\alpha_c \in [-\infty, \infty]$ such that

$$m_\xi(\alpha, A) = 0, \quad \alpha > \alpha_c \quad \text{if } \alpha_c \neq +\infty,$$
$$m_\xi(\alpha, A) = \infty, \quad \alpha < \alpha_c \quad \text{if } \alpha_c \neq -\infty.$$

Table 4.1

Generalized Carathéodory Construction	
X-space $\equiv$ set	Example Metric Space
$\mathcal{F}$-collection of subsets	Open Subsets; Balls $B_\varepsilon(x) = \{y\colon \operatorname{dist}(x, y) < \varepsilon\}$
$\psi : \mathcal{F} \to \mathbb{R}^+$ – a function, s.t. Axiom A: For any $\varepsilon > 0$ there is a finite or countable subcollection $\{U_i\}$ with $\Psi(U_i) \leqslant \varepsilon$, s.t. $\bigcup_i U_i \supset X$.	$\psi(U) = \operatorname{diam}(U)$ $\operatorname{diam} U_i \leqslant \varepsilon$
$\xi, \eta : \mathcal{F} \to \mathbb{R}^+$ are functions s.t. Axiom B: $\begin{cases} \xi(U) \geqslant 0, & \forall U \in \mathcal{F} \\ \eta(U) > 0, & \forall U \neq \\ \emptyset,\ U \in \mathcal{F} \end{cases}$	Hausdorff $\xi(U) \equiv 1$ $\eta(U) = \operatorname{diam}(U)$
Axiom C: For any $\delta < 0\ \exists \varepsilon > 0$ such that $\eta(U) \leqslant \delta$ provided that $\psi(U) \leqslant \varepsilon$ (for *any* $U \in \mathcal{F}$).	Carathéodory $\xi(U) \equiv 1$ $\eta(U) = \phi(\operatorname{diam} U)$

The number α_c is said to be the *Carathéodory dimension* of set A relative to the structure $(\mathcal{F}, \psi, \xi, \eta)$ – see Table 4.2.

The foundation of the theory of fractal dimension, now known as Hausdorff dimension, was established by Carathéodory and Hausdorff. In [40] Carathéodory shows how to define a p-dimensional measure for sets in q-dimensional spaces. Then, in [64], Hausdorff adapts the definition so that it makes sense even if p is not an integer. It is very interesting to see also a function of the diameter that is included in the definition of Hausdorff's measure.

4.1.1. Examples (see Table 4.3)

- If $\mathcal{F}$ is the collection of all open balls $\{B(x, \varepsilon)\}$ of all diameters $\varepsilon > 0$ centered at all points $x \in X$, $\xi(B(x, \varepsilon)) \equiv 1$, $\eta(B(x, \varepsilon)) = \varepsilon$, then $\alpha_c = \dim_H A$, the Hausdorff dimension of set A.
- A nontrivial example is the dimension-like definition of the topological entropy [97]. Assume that X is compact and $f : X \to X$ is a continuous map. Given $n > 0$, $\varepsilon_0 > 0$, the *Bowen set* is defined as

$$B_n(x, \varepsilon_0) = \{y \in X\colon \rho(f^i x, f^i y) \leqslant \varepsilon_0,\ 0 \leqslant i \leqslant n\}.$$

Let $\mathcal{F}$ be the set of all Bowen sets. Let

$$\xi(B_n(x, \varepsilon_0)) \equiv 1,\ \psi(B_n(x, \varepsilon_0)) = 1/n \text{ and } \eta(B_n(x, \varepsilon_0)) = \mathrm{e}^{-n}.$$

Table 4.2

Carathéodory dimension	
$A \subset X$	Example
$M_\xi(\alpha, \varepsilon, A) = \inf\limits_{\substack{G \\ \psi(u_i)\leqslant\varepsilon}} \Big\{ \sum\limits_{\substack{\bigcup u_i \supset A \\ u_i \in G \\ \psi(u_i)\leqslant\varepsilon}} \xi(u_i)\cdot\eta(u_i)^\alpha \Big\}$ $M_\xi(\alpha, \varepsilon, A) \nearrow$ if $\varepsilon \searrow \quad \Rightarrow$ $\exists \lim\limits_{\varepsilon\to 0} M_\xi(\alpha, \varepsilon, A) = m_\xi(\alpha, A)$	$\inf\limits_G \Big\{ \sum\limits_{\operatorname{diam} u_i \leqslant \varepsilon} \operatorname{diam} u_i^\alpha \Big\}$ or $\inf\limits_G \Big\{ \sum\limits_{\operatorname{diam} u_i \leqslant \varepsilon} \phi(\operatorname{diam} u_i)^\alpha \Big\}$
Properties of m_ξ as a function of A (i) $A_1 \subset A_2 \Rightarrow m_\xi(\alpha, A_1) \leqslant m_\xi(A_2, \alpha)$ (ii) $m_\xi(\bigcup_k A_k, \alpha) \leqslant \sum_k m_\xi(\alpha, A_k)$ If $m_\xi(\alpha, \emptyset) = 0$ then m_ξ is outer measure.	$m_H(A, \alpha)$ outer Hausdorff measure
$m_\xi(\alpha, A)$ as a function of α 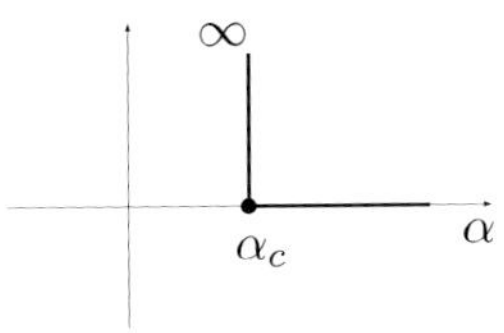 Carathéodory dimension of A $\alpha_c = \sup\{\alpha\colon m_\xi(\alpha, A) = \infty\}$ $= \inf\{\alpha\colon m_\xi(\alpha, A) = 0\}$ $= \dim_\xi(A)$	$\dim_H(A)$ Hausdorff dimension

Then the ε_0-topological entropy of f on A, $h_{\text{top}}(f|A, \varepsilon_0)$, is the Carathéodory dimension α_c and

$$h_{\text{top}}(f|A) := \limsup_{\varepsilon_0\to 0} h_{\text{top}}(f|A, \varepsilon_0).$$

It was shown in [97] that this entropy coincides with the standard topological entropy if A is a compact and f-invariant set.

- If we consider the set function

$$\xi\big(B_n(x,\varepsilon)\big) = \exp\left(\sup_{y\in B_n(x,\varepsilon)} \sum_{k=0}^{n-1} \phi\big(f^k y\big) \right), \tag{4.2}$$

Table 4.3

Examples				
$\mathcal{F}$	ξ	ψ	η	Result
Open sets U (or balls)	1	diam U	diam U	$\dim_H(A)$
Bowen sets $B_n(x,\varepsilon)$	1	$\frac{1}{n}$	e^{-n}	h_{top} if $A \subset X$ is inv. and compact
same	Eq. (4.2)	same	same	$P(\phi)$ – top pressure if $A \subset X$ is inv. and compact
Balls $B_\varepsilon(x)$	$\mu(B_\varepsilon(x))^q$	ε	ε	$\dim_q(\mu)$ (if μ is regular) including $HP_q(\mu)$ and Correlation dimension
Open sets U	$e^{-q\tau(U)}$	diam U	diam U	Dimension for Poincaré recurrences

$\psi(B_n(x,\varepsilon)) = 1/n$, $\eta(B_n(x,\varepsilon)) = \exp(-n)$, where $\phi : X \to \mathbb{R}$ is continuous, then we obtain the Carathéodory dimension $\alpha_c = P(\phi)$ which is called the Carathéodory topological pressure of the "potential" ϕ. It was shown in [97] that if A is invariant and compact then α_c coincides with the standard topological pressure.

- One can study dimension-like characteristics of a measure μ by using the generalized Carathéodory construction. For example, if one chooses

$$\xi\big(B_\varepsilon(x)\big) = \mu\big(B_\varepsilon(x)\big)^q \quad \text{and} \quad \psi\big(B_\varepsilon(x)\big) = \eta\big(B_\varepsilon(x)\big) = \varepsilon,$$

  then one arrives to a characteristic $\dim_q(\mu)$ of the measure μ that is very similar to so called Hentschel–Procaccia spectrum $HP_q(\mu)$, see [97]. One can show also that the Billingsley dimension [24,25] can be expressed in terms of the Carathéodory dimension.
- We define below the Poincaré recurrence $\tau(U)$ of the set U. If we choose $\exp(-q\tau(U))$ in the capacity of the "gauge function" ξ, and

$$\psi(U) = \eta(U) = \operatorname{diam} U$$

  we obtain the Carathéodory dimension which we call the spectrum of dimensions for Poincaré recurrences. In fact, the book is devoted to the study of this dimension.

4.2. The spectrum of dimensions for recurrences

Typical orbits in Hamiltonian systems and orbits in attractors in dissipative systems repeat their behavior in time. This repetition can be expressed in terms of Poincaré recurrences.

Consider a dynamical system (X, f) where X is a metric space and the mapping $f : X \to X$ is continuous. Let $A \subset X$ be a f-invariant subset. In the framework of the general Carathéodory construction we consider covers by open balls. For each $A \subset X$, denote by $\mathcal{B}_\varepsilon(A)$ the class of all finite or countable covers of A by balls of diameter less than or equal to ε. For an open ball $B \subset X$ let the *Poincaré recurrence* be defined as

$$\tau(B) = \inf\{\tau(x, B)\colon\ x \in B\},$$

where $\tau(x, B) = \min\{t \geqslant 1\colon\ f^t(x) \in B\}$ is the first return time of $x \in B$. Given $G \in \mathcal{B}_\varepsilon(A)$ and $\alpha, q \in \mathbb{R}$, consider the sum

$$M_\xi(\alpha, q, \varepsilon, G, A) = \sum_{B \in G} \xi\big(\tau(B)\big)^q \operatorname{diam} B^\alpha, \tag{4.3}$$

where the real non-negative gauge function $\xi : R \to \mathbb{R}$ is such that $\xi(t) \to 0$ as $t \to \infty$. Below we will consider the functions $\xi(t) = e^{-t}$ and $\xi(t) = 1/t$.

Next we define

$$M_\xi(\alpha, q, \varepsilon, A) = \inf\{M_\xi(\alpha, q, \varepsilon, G, A)\colon\ G \in \mathcal{B}_\varepsilon(A)\}. \tag{4.4}$$

For fixed q the limit $m_\xi(\alpha, q, A) = \lim_{\varepsilon \to 0} M_\xi(\alpha, q, \varepsilon, A)$ has an abrupt change from infinity to zero as one varies α from minus infinity to infinity. There is a unique critical value

$$\alpha_c(q, \xi, A) = \sup\{\alpha\colon\ m_\xi(\alpha, q, A) = \infty\} \tag{4.5}$$

such that $m_\xi(\alpha, q, A) = \infty$ if $\alpha < \alpha_c(q, \xi, A)$, provided that $\alpha_c(q, \xi, A) \neq -\infty$, and $m_\xi(\alpha, q, A) = 0$ if $\alpha > \alpha_c(q, \xi, A)$, provided that $\alpha_c(q, \xi, A) \neq \infty$.

The function $\alpha_c(q) := \alpha_c(q, \xi, A)$ is said to be the *spectrum of dimensions for Poincaré recurrences*, specified by the function ξ. The quantity

$$q_0 = \sup\{q\colon\ \alpha_c(q) > 0\}$$

has been introduced in [9,10] and [11]. It is said to be the *dimension for Poincaré recurrences* of set $A \subset X$ specified by the gauge function ξ. In the case $\xi(t) = \exp(-t)$, a quantity similar to q_0 was introduced in [95] and was called the AP-dimension. Roughly speaking, q_0 is the smallest solution of the equation $\alpha_c(q, \xi, A) = 0$.

Not many specific examples are known where the dimension for Poincaré recurrences has been explicitly computed or estimated ([36,76,78]).

4.3. Dimension and capacities

We proceed in a slightly different way than in Section 4.2 to introduce the dimension for Poincaré recurrences.

Fix an arbitrary monotonically decreasing function $\xi(t)$, $t > 0$, $\lim_{t\to\infty} \xi(t) = 0$ and introduce the following structure: $\mathcal{F}$ is the collection of all open sets in X; $\tau(u)$ is the Poincaré recurrence for the set $u \in \mathcal{F}$. Then, for a compact $A \subset X$, consider the quantities

$$M_\xi(0, q, \varepsilon, A) = \inf_G \sum_i \xi\big(\tau(u_i)\big)^q,$$

where infimum is taken over all covers $G = \{u_i\}$, $\operatorname{diam} u_i \leqslant \varepsilon$, and

$$R_\xi(0, q, \varepsilon, A) = \inf_H \sum_i \xi\big(\tau(u_i)\big)^q,$$

where infimum is taken over all covers $H = \{u_i\}$, $\operatorname{diam} u_i = \varepsilon$.

DEFINITION 4.1. Since $\tau(u_i)$ is nondecreasing as $\varepsilon \to 0$ and $\xi(t)$ is monotone, then there exists the limit

$$\lim_{\varepsilon\to 0} M_\xi(0, q, \varepsilon, A) =: m_\xi(q, A).$$

Since $\xi \circ \tau(u_i) \to 0$ as $\varepsilon \to 0$ then

$$m_\xi(q, A) = \infty \quad \text{if } q < 0.$$

Let

$$q_* := \sup\big\{q\colon m_\xi(q, A) = \infty\big\}.$$

The critical value q_* is said to be the dimension for Poincaré recurrences of set A, specified by the function ξ.
Consider next

$$\overline{m}_\xi(q, A) = \limsup_{\varepsilon\to 0} R(0, q, \varepsilon, A) \quad \text{and} \quad \underline{m}_\xi(q, A) = \liminf_{\varepsilon\to 0} R(0, q, \varepsilon, A).$$

Again, $\overline{m}_\xi(q, A) = \infty$ if $q < 0$. Let $\bar{q}_0 = \sup\{q\colon \overline{m}_\xi(q, A)\}$. The upper $\bar{q}_0$ and lower $\underline{q}_0$ capacities are said to be the upper and lower capacities of set A, specified by the function ξ. We denote them as follows: $q_* = \dim_\xi(A)$, $\bar{q}_0 = \overline{\dim}_\xi(A)$, $\underline{q}_0 = \underline{\dim}_\xi(A)$.

The following result was established in [95].

LEMMA 4.1. *Assume* $\alpha_c(q) > 0$ *for* $q \in [0, q_*)$. *Assume that* $\lim_{q\nearrow q_*} \alpha_c(q) = 0$. *Then* $q_0 = q_*$.

It follows from the general theory [97–99] that

$$\dim_\xi(A) \leqslant \underline{\dim}_\xi(A) \leqslant \overline{\dim}_\xi(A)$$

for any admissible function $\xi(t)$.

In order to understand the meaning of these notions, let us imagine a nice situation for which

$$\dim_\xi(A) = \overline{\dim}_\xi(A) = \underline{\dim}_\xi(A) = q_0, \quad 0 < q_0 < \infty,$$

and $\underline{\dim}_B A = \overline{\dim}_B A = b$. Then, we have the following estimate for the average

$$\left\langle \xi\left(\tau(u_i)\right)^{q_0}\right\rangle = \frac{1}{N}\sum_{i=1}^{N} \xi\left(\tau(u_i)\right)^{q_0} \sim \varepsilon^b, \quad \varepsilon \ll 1,$$

over a cover $H = \{u_i\}$ with $\operatorname{diam} u_i = \varepsilon$. This estimation tells us that we can expect the average value of $\tau(u_i)$ to behave as follows:

$$\left\langle \tau(u_i)\right\rangle \sim \xi^{-1}\left(\varepsilon^{b/q_0}\right).$$

For a gauge function $\xi(t) \sim 1/t$ we have

$$\left\langle \tau(u_i)\right\rangle \sim \varepsilon^{-b/q_0}, \quad \varepsilon \ll 1. \tag{4.6}$$

Similarly, if $\xi(t) \sim e^{-t}$, then

$$\left\langle \tau(u_i)\right\rangle \sim -\frac{b}{q_0}\log\varepsilon, \quad \varepsilon \ll 1, \tag{4.7}$$

and so forth. In a general case we may obtain some estimates (from above and from below) for Poincaré recurrences with the help of the quantities $\overline{\dim}_\xi(X)$ and $\underline{\dim}_\xi(X)$ provided that they are finite nonzero numbers.

4.4. The appropriate gauge functions

It is possible to assume that the value of $\underline{\dim}_\xi(X)$ (or $\overline{\dim}_\xi(X)$) is equal to zero (or infinity). If so, we should change the function ξ and find a suitable one. The existence of such suitable gauge functions can be proved under the assumption that the lowest upper bound for recurrences

$$\overline{\tau}(\varepsilon) = \sup\left\{\tau\left(B(x,\varepsilon)\right) \colon x \in X\right\}$$

is finite, $\overline{\tau}(\varepsilon) < \infty$. This is a very general assumption that implies that every point in X is non-wandering. Minimal sets are examples of such a situation, i.e., the quantity $\overline{\tau}(\varepsilon)$ is finite for minimal X. Otherwise, due to compactness of X, we could find a sequence of points $x_n \to x_*$ as $n \to \infty$ with

$$\tau\left(B(x_n,\varepsilon)\right) \geqslant n. \tag{4.8}$$

However, consider the following remarks.

1. The discreteness of time implies that for any point $x_* \in X$ there is a point $y \in B(x_*, \varepsilon)$ such that

$$\tau\bigl(B(x_*, \varepsilon)\bigr) = \tau\bigl(y, B(x_*, \varepsilon)\bigr).$$

2. The openness of $B(x_*, \varepsilon)$ implies that

$$\rho(y, x_*) < \varepsilon \quad \text{and} \quad \rho\bigl(f^{\tau(y, B(x_*, \varepsilon))} y, x_*\bigr) < \varepsilon. \tag{4.9}$$

3. The assumption that $\lim_{n\to\infty} \rho(x_n, x_*) = 0$ together with the inequalities (4.9) imply that the points y and $f^{\tau(y, B(x_*, \varepsilon))} y$ belong to the ball $B(x_n, \varepsilon)$ for every sufficiently large n, say, $n \geqslant n_0$.

Thus,

$$\tau\bigl(B(x_n, \varepsilon)\bigr) = \inf_{z\in B(x_n,\varepsilon)} \tau\bigl(z, B(x_n, \varepsilon)\bigr) \leqslant \tau\bigl(y, B(x_*, \varepsilon)\bigr).$$

This inequality contradicts (4.8), so $\overline{\tau}(\varepsilon) < \infty$ for minimal X.

The following statement tells us that an appropriate gauge function exists.

THEOREM 4.1. *Assume that $\overline{\tau}(\varepsilon) < \infty$ in X. Then the following statements hold.*

(1) *If $\overline{\dim}_B X < \infty$ then there exists a function $\xi_1(t)$, $t \geqslant t_0$, $\lim_{t\to\infty} \xi_1(t) = 0$, such that $\overline{\dim}_{\xi_1}(X) \leqslant \overline{\dim}_B X$.*

(2) *If $\underline{\dim}_B X > 0$ then, for $t \geqslant t_0$, there exists a function $\xi_2(t)$, with $\lim_{t\to\infty} \xi_2(t) = 0$ and such that $\underline{\dim}_{\xi_2}(X) \geqslant \underline{\dim}_B X$.*

PROOF. First, we prove statement (2). Let $\varepsilon_1 \geqslant \varepsilon_2$ so that $\tau(B(x, \varepsilon_1)) \leqslant \tau(B(x, \varepsilon_2))$ for every point $x \in X$. Therefore, $\overline{\tau}(\varepsilon_1) \leqslant \overline{\tau}(\varepsilon_2)$.

Assume first that $\overline{\tau}(\varepsilon) \to \infty$ as $\varepsilon \to 0$. Hence, $\overline{\tau}(\varepsilon)$ has a countable number of points of jump discontinuity on any interval $(0, \varepsilon_0]$, $\varepsilon_0 > 0$. Let $\Lambda \subset (0, \varepsilon_0]$ be the set of points of jump discontinuity of the function $\overline{\tau}(\varepsilon)$.

Define a function $\xi_2(t)$ as follows. For values of

$$t \in \overline{\tau}(\Lambda) := \{t_1, \dots, t_n, t_{n+1}, \dots\} \subset \mathbb{Z}_+$$

we set

$$\xi_2\bigl(\overline{\tau}(\varepsilon_*)\bigr) = \varepsilon_*, \quad \varepsilon_* \in \Lambda, \quad \varepsilon_*(t) = \sup\bigl\{\varepsilon \notin \Lambda,\ \overline{\tau}(\varepsilon) = t\bigr\}.$$

For values of $t \in (t_n, t_{n+1})$, where n is an integer, set $\xi_2(t) = At + B$ where

$$A = \frac{\xi_2(t_{n+1}) - \xi_2(t_n)}{t_{n+1} - t_n}, \quad \text{and} \quad B = \frac{t_{n+1}\xi_2(t_n) - t_n\xi_2(t_{n+1})}{t_{n+1} - t_n}.$$

The constructed function $\xi_2(t)$ is continuous and has the desired properties: $\lim_{t\to\infty} \xi_2(t) = 0$; $\xi_2(t) \leqslant \xi_2(t')$ if $t \leqslant t'$; $\xi_2(\overline{\tau}(\varepsilon)) \leqslant \varepsilon$.

Given $\varepsilon > 0$ consider a finite cover G of X by open balls $B(x_i, \varepsilon)$, $i = 1, \ldots, N(\varepsilon)$. Since $\tau(B(x_i, \varepsilon)) \leqslant \overline{\tau}(\varepsilon)$, then $\xi_2(\tau(B(x_i, \varepsilon))) \geqslant \xi_2(\overline{\tau}(\varepsilon)) \geqslant \varepsilon$. Therefore, for any $b > 0$ we have

$$R_{\xi_2}(0, \underline{q} - b, \varepsilon, \xi_2) = \sum_{i=1}^{N} \xi_2\big(\tau\big(B(x_i, \varepsilon)\big)\big)^{\underline{q}-b} \geqslant N\varepsilon^{\underline{q}-b}, \tag{4.10}$$

where $\underline{q} = \underline{\dim}_B X$. It follows from the definition of the lower box dimension that for an arbitrary large constant C there is $\varepsilon_0 > 0$ such that $N\varepsilon^{\underline{q}-b} > C$ if $0 < \varepsilon < \varepsilon_0$. Thus, $R(\underline{q}-b, \varepsilon, \xi_2) > C$ if ε is small enough and $\underline{\dim}_{\xi_2}(X) \geqslant \underline{q}-b$. Thanks to the arbitrariness of b, we obtain the desired result.

The proof of statement (1) of the theorem is the same. We substitute $\underline{\tau}(\varepsilon) = \inf\{\tau(B(x, \varepsilon))\colon x \in X\}$ instead of $\overline{\tau}(\varepsilon)$ and omit the first part of the proof since $\underline{\tau}(\varepsilon) < \infty$ by definition. In a way identical to the one for $\overline{\tau}(\varepsilon)$, we show that $\underline{\tau}(\varepsilon)$ is a monotone function. Moreover, it follows from the definition of $\underline{\tau}(\varepsilon)$ that $\lim_{\varepsilon \to 0} \underline{\tau}(\varepsilon) = \infty$. We denote by $\Lambda \subset (0, \varepsilon_0]$ the set of points of jump discontinuity of the function $\underline{\tau}(\varepsilon)$ and define $\xi_1(t)$, $t \in \underline{\tau}(\Lambda) \subset \mathbb{Z}_+$, by setting:

1. $\xi_1(\underline{\tau}(\varepsilon_*)) \equiv \varepsilon_*$, $\varepsilon_* \in \Lambda$, $\varepsilon_*(t) = \inf\{\varepsilon \notin \Lambda, \underline{\tau}(\varepsilon) = t\}$.
2. $\xi_1(t)$ is defined in the same way as $\xi_2(t)$ for other values of t.

It follows from the definition that $\xi_1(\underline{\tau}(\varepsilon)) \leqslant \varepsilon$. Moreover,

$$\xi_1(t) \leqslant \xi_2(t). \tag{4.11}$$

Let $\overline{q} = \overline{\dim}_B X < \infty$. Then, given $\varepsilon > 0$ consider a finite cover G of X by open balls $B(x_i, \varepsilon)$, $i = 1, \ldots, N = N(\varepsilon)$. By definition of $\xi_1(t)$, we have that

$$\xi_1\big(\tau\big(B(x_i, \varepsilon)\big)\big) \leqslant \xi_1\big(\underline{\tau}(\varepsilon)\big) \leqslant \varepsilon.$$

Hence, for any $b > 0$

$$\sum_{i=1}^{N} \xi_1\big(\tau\big(B(x_i, \varepsilon)\big)\big)^{\overline{q}+b} \leqslant N(\varepsilon)\varepsilon^{\overline{q}+b} \leqslant 1 \tag{4.12}$$

(by definition of the upper box dimension $N(\varepsilon)\varepsilon^{\overline{q}+b} \leqslant 1$, $\varepsilon \ll 1$). Therefore $\overline{\dim}_{\xi_1}(X) \leqslant \overline{q} + b$. Since b is chosen to be arbitrary small, then $\overline{\dim}_{\xi_1}(X) \leqslant \overline{q}$. □

Remark that if X contains wandering points, then the condition $\overline{\tau}(\varepsilon) < \infty$ in Theorem 4.1 could not be satisfied. In this situation we can restrict ourselves to the center set of the dynamical systems, which contains no wandering points.

The domain of definition of the gauge functions constructed in the proof of Theorem 4.1 is $[0, \infty)$ whenever $\overline{\tau}(\varepsilon) \to \infty$ as $\varepsilon \to 0$. This would not be the case if for every point $x \in X$ and every $\varepsilon > 0$ the ball $B(x, \varepsilon)$ contains a periodic point of period at most $T < \infty$ since then $\lim_{\varepsilon \to 0} \overline{\tau}(\varepsilon) \to T < \infty$ as $\varepsilon \to 0$.

However, in this situation the proof of Theorem 4.1 is valid if the gauge function is defined, e.g., as $\underline{\varepsilon}\exp(T-t)$ for $t \in [T, \infty)$, with $\underline{\varepsilon} := \inf\{\varepsilon\colon \overline{\tau}(\varepsilon) < T\}$.

We note that the functions ξ_1 and ξ_2 (constructed above) satisfy the inequality $\xi_1(t) \leqslant \xi_2(t)$, $t \geqslant t_1$ (see Eq. (4.11)). Therefore, $\dim_{\xi_1}(X) \leqslant \dim_{\xi_2}(X)$. It is not enough to claim the existence of just one function servicing both lower and upper capacities. We need an additional condition for that.

COROLLARY 4.1. *Under the conditions of Theorem* 4.1, *there exists a function* $\xi(t)$, $\lim_{t\to\infty}\xi(t) = 0$, *such that* $0 < \underline{\dim}_{\xi}(X) \leqslant \overline{\dim}_{\xi}(X) < \infty$ *if*

$$\lim_{t\to\infty}\frac{\xi_2(t)}{\xi_1(t)} < \infty \tag{4.13}$$

where $\xi_{1,2}$ *are functions constructed in the proof of Theorem* 4.1.

PROOF. The inequality (4.13) implies that

$$\xi_1(t) \geqslant C\xi_2(t) \tag{4.14}$$

if $t \geqslant t_1 > 0$, and C is constant. Define $\xi(t) = \xi_1(t)\xi_2(t)$ where ξ_1, ξ_2 are functions constructed in the proof of Theorem 4.1. Given $\varepsilon > 0$, consider a finite cover G by balls $B(x_i, \varepsilon)$, $i = 1, \ldots, N$. The condition (4.14) implies (for $\varepsilon \ll 1$) that

$$\sum_{i=1}^{N} C^q\xi_2^{2q}\big(\tau\big(B(x_i,\varepsilon)\big)\big) \leqslant \sum_{i=1}^{N}\xi^q\big(\underline{\tau}\big(B(x_i,\varepsilon)\big)\big) \leqslant \sum_{i=1}^{N}\frac{1}{C^q}\xi_1^{2q}\big(\tau\big(B(x_i,\varepsilon)\big)\big),$$

and the desired result follows from Eqs. (4.10) and (4.12). □

The assumption (4.13) can be represented in another form. The inequalities (4.11) and (4.14), i.e.,

$$C\xi_2(t) \leqslant \xi_1(t) \leqslant \xi_2(t), \quad t \geqslant T > 0, \tag{4.15}$$

indicate the requirement to have similar asymptotic behavior as $t \to \infty$. In the proof of Theorem 4.1 we introduced the quantities

$$\underline{\tau}(\varepsilon) = \inf_{x\in M}\tau\big(B(x,\varepsilon)\big) \quad \text{and} \quad \overline{\tau}(\varepsilon) = \sup_{x\in M}\tau\big(B(x,\varepsilon)\big),$$

so that $\xi_1(\underline{\tau}(\varepsilon)) \leqslant \varepsilon$, $\xi_2(\overline{\tau}(\varepsilon)) \geqslant \varepsilon$, for almost every $\varepsilon \in (0, \varepsilon_0]$. The condition (4.15) can thus be reflected in some properties of $\underline{\tau}(\varepsilon)$, $\overline{\tau}(\varepsilon)$. For example, if $\xi_1(t) \sim A/t^\beta$, $\xi_2(t) \sim B/t^\beta$, $t \gg 1$ (A, B, and β are constants), then

$$\overline{\lim_{\varepsilon\to 0}}\frac{\log\overline{\tau}(\varepsilon)}{\log\underline{\tau}(\varepsilon)} < \infty;$$

if $\xi_1(t) \sim Ae^{-\beta t}$, $\xi_2(t) \sim Be^{-\beta t}$, $t \gg 1$, then

$$\overline{\lim_{\varepsilon \to 0}} \frac{\overline{\tau}(\varepsilon)}{\underline{\tau}(\varepsilon)} < \infty,$$

etc. Thus, the maximal and minimal Poincaré recurrences for any fixed small ε should be of the same order (in a corresponding scale).

4.5. General properties of the dimension for recurrences

Specific properties of q_0 are sensible to the kind of gauge function ξ being used. Originally the dimension for Poincaré recurrences was introduced to deal with irrational rotations on the circle [9]. By using the gauge function $\xi(t) = 1/t$ the dimension q_0 happened to be related to the rate of Diophantine approximation of the rotation number. These results are discussed in Section 4.6.2 below. Generally, for systems with zero topological entropy a good choice of gauge function is $\xi(t) = 1/t$, while for positive entropy systems a good choice is $\xi(t) = \exp(-t)$.

The dimension for Poincaré recurrences, when defined over covers by open sets shares with the topological entropy the property of being a topological invariant. If the dynamical system (X, f), on compact X, is generated by a continuous map f, then the dimension for Poincaré recurrences q_0 is preserved by uniform homeomorphisms.

Let (X, d) and (X', d') be two metric spaces. A homeomorphism $h : X \to X'$ is said to be uniform if for every $\delta > 0$ there exists ε_δ (which goes to 0 as $\delta \to 0$) such that for any pair of points $x, y \in X$ with distance $d(x, y) < \varepsilon_\delta$, the inequality $d'(h(x), h(y)) < \delta$ holds. Two dynamical systems (X, f) and (X', f') are uniformly homeomorphic if there exists a uniform homeomorphism $h : X \to X'$ such that $f' \circ h = h \circ f$.

THEOREM 4.2. *Let (X, f) and (X', f') be uniformly homeomorphic with homeomorphism $h : X \to X'$. Then, for each $A \subset X$ and $A' = h(A) \subset X'$, $q_0(A) = q_0'(A')$.*

PROOF. Let $\mathcal{B}'_\delta(A')$ be the family of all coverings G' of subset A' with open sets of diameter not greater than $\delta > 0$. Let ε_δ be such that $d(x, y) < \varepsilon_\delta$ implies $d'(h(x), h(y)) < \delta$. Let $h(\mathcal{B}_{\varepsilon_\delta}(A))$ be the set of coverings of A of diameter not greater than ε_δ pushed forward to X', with elements $h(G) = \{h(u) \colon u \in G\}$, for each $G \in \mathcal{B}_{\varepsilon_\delta}(A)$. Thus $\mathcal{B}'_\delta(A') \supset h(\mathcal{B}_{\varepsilon_\delta}(A))$ and then

$$\begin{aligned} M_\xi(0, q, \varepsilon_\delta, A) &= \inf_{G \in \mathcal{B}_{\varepsilon_\delta}(A)} \sum_{u \in G} \xi\big(\tau(u)\big)^q \\ &\geqslant \inf_{G \in \mathcal{B}'_\delta(A')} \sum_{u \in G} \xi\big(\tau'(u)\big)^q = M'_\xi(0, q, \delta, A') \end{aligned}$$

where $\tau' = \tau \circ h^{-1}$ is the Poincaré recurrence for the primed system. By interchanging the roles of systems (X, f) and (X', f') in the preceding argument we conclude that

$$M_\xi(0, q, \varepsilon_\delta, A) = M'_\xi(0, q, \delta, A').$$

So, q_0 is a topological invariant. □

Further similarities of entropy with the q_0 dimension are that $q_0(X)$ coincides with the dimension of the non-wandering set (when f is restricted to it) and that the set of periodic points of system (X, f) provide an essential contribution to the dimension of Poincaré recurrences. This is mainly due to the fact that if $x \in X$ is periodic with smallest period n, then $\tau(u) \leqslant n$ for any $u \ni x$, irrespective of its diameter. Thus, periodic points put a natural lower bound to the dimension for Poincaré recurrences.

THEOREM 4.3. (See [95].) *The dimension for Poincaré recurrences has the following properties.*

(1) $q_0(X, f) = q_0(NW, f) = q_0(NW, f|NW)$, *where* NW *denotes the set of non-wandering points.*
(2) *If* $\xi(t) = \exp(-t)$, *and if the number of periodic points of smallest period k is finite for every k, then*

$$q_0 \geqslant \limsup_{n\to\infty} \frac{1}{n} \log \#\{x\colon\ f^n(x) = x\}.$$

(3) *If* $\xi(t) = \exp(-t)$, *then for any* $k > 0$, *we have that* $q_0(X, f^k) \leqslant kq_0(X, f)$.
(4) *If* $\xi(t) = 1/t$, *then for any* $k > 0$, *we have that* $q_0(X, f^k) \leqslant q_0(X, f)$.

In many systems the limit in the statement (2) of Theorem 4.3 coincides with the topological entropy $h_{top}(X)$ and the equality sign holds. For them $h_{top}(\Omega) = q_0(\Omega)$, and the dimension for Poincaré recurrences is not a new topological invariant. Subshifts of finite type are examples of such systems.

THEOREM 4.4. *Let* (Ω, σ) *be a subshift of finite type, with finite or infinite alphabet and such that* $\#\{x\colon\ f^n(x) = x\} < \infty$ *for every n. Then*

$$q_0(\Omega) = \limsup_{n\to\infty} \frac{1}{n} \#\{x\colon\ f^n(x) = x\}.$$

PROOF. Consider the particular cover of Ω by n-cylinders. In this cover the number of cylinders with first return time k is at most the number of periodic points with period k. Thus,

$$M_\xi(0, q, \varepsilon, \Omega) \leqslant \sum_k \#\mathrm{Per}(k)\exp(-qk)$$

where $\xi(t) = \exp(-t)$. The upper bound remains true in the $\varepsilon \to 0$ limit. The lower bound is provided by statement 2 of Theorem 4.3. □

In [36] minimal subshifts of positive entropy are constructed for which $0 < q_0 < h_{\mathrm{top}}(\Omega)$ with gauge function $\xi(t) = \exp(-t)$.

THEOREM 4.5. (See [36].) *For each subshift $\Omega \subset \Omega_2$, $q_0(\Omega) \leqslant h_{\mathrm{top}}(\Omega)$. There exist subshifts $\Omega \subset \Omega_2$ for which $q_0(\Omega) \neq h_{\mathrm{top}}(\Omega)$.*

4.6. Dimension for Poincaré recurrences for minimal sets

Let X be a metric space with metric ρ and let $f : X \to X$ be a continuous map. For example, X is an invariant torus (circle) or cantorus in the phase space of a conservative system generated by a map, f is the restriction of the map to X, and ρ is the restriction of a distance in the phase space to the points of X. The set X is said to be minimal if it is closed, invariant and does not contain another closed invariant subset. Minimal sets have the following nice properties:

1. For any $\varepsilon > 0$ there is an integer number $N = N(\varepsilon)$ such that

   $$\bigcup_{k=0}^{N-1} B\big(f^k x, \varepsilon\big) \supset X \quad \text{for any } x \in X.$$

   Here $B(y, \varepsilon)$ is the ball of diameter ε centered at point y. Thus, the minimal set X can be approximated by the finite piece of any orbit with an arbitrary accuracy ε.
2. For any $x \in X$, the orbit $\bigcup_{k=0}^{\infty} f^k x$ is everywhere dense in X, i.e., X is the closure of any (recurrent) orbit in it.

Of course, a periodic orbit is a (trivial) minimal set. From now on, we will consider nontrivial compact minimal sets.

A Birkhoff theorem tells us that the closure of a recurrent orbit contains uncountably many recurrent orbits (see, e.g., Ref. [89]), so nontrivial minimal sets are formed by many orbits with nontrivial behavior. Let us also remark that from the ergodic theory viewpoint minimal sets and ergodic systems are very similar subjects: a Jewett–Krieger theorem implies that every ergodic invertible map of a Lebesgue space is measurably isomorphic to a minimal uniquely ergodic homeomorphism of a zero-dimensional compact metric space (see, e.g., Ref. [120]).

4.6.1. The gauge function $\xi(t) = 1/t$

For some minimal sets a good choice of gauge function is $\xi(t) = 1/t$ (see the examples in the following sections). In this case, Kac's theorem (see Chapter 17) puts a lower bound for the dimension for Poincaré recurrences.

THEOREM 4.6. (See [95].) *Let (X, f) be a minimal dynamical system, having a Borel ergodic measure μ. Let $A \subset X$ have positive measure, $\mu(A) > 0$. Then*

$$\dim_{(1/t)}(A) \geqslant 1.$$

Remark that the theorem implies that $\dim_{(1/t)}(X) \geqslant 1$. (4.16)

PROOF. By Kac's theorem we have that

$$\tau(u) \leqslant \frac{1}{\mu(u)} = \int \tau(x, u) \frac{\mu(dx)}{\mu(u)}.$$

Then, for an arbitrary cover G of $A \subset X$ with open sets u of diameter less than ε we may write

$$M_{(1/t)}(0, 1, \varepsilon, G, A) = \sum_{u \in G} \frac{1}{\tau(u)} \geqslant \sum_{u \in G} \mu(u) \geqslant \mu(A) > 0.$$

Since the system is assumed to be minimal (it has no periodic points), this proves that $\dim_{(1/t)}(A) \geqslant 1$. □

In the following sections we consider several examples of minimal sets.

4.6.2. Rotations of the circle

The simplest minimal set is the circle $S^1 = \{x \pmod 1\} = \mathbb{R}/\mathbb{Z}$ and the simplest recurrent orbits are the orbits generated by the rigid rotation $f = f_\alpha : x \mapsto x + \alpha \pmod 1$ where α is an irrational number. It is well known that α can be approximated by rational numbers m/n (m and n are relatively prime) such that

$$\left|\alpha - \frac{m}{n}\right| < n^{-\nu-1} \tag{4.17}$$

for some value ν and some pairs (m, n). Let $\nu(\alpha) = \sup\{\nu\}$ where the supremum is taken over all ν for which the inequality (4.17) has infinitely many solutions (m, n) with $n > 0$. In other words, if $\nu(\alpha)$ is finite, then for positive $\delta \ll 1$ the inequality

$$\left|\alpha - \frac{m}{n}\right| \geqslant n^{-\nu(\alpha)-1-\delta} \tag{4.18}$$

holds except (possibly) for a finite number of relatively prime pairs (m, n).

We establish a relationship between dimensions for recurrences for rotations on the circle and the rate $\nu(\alpha)$ of Diophantine approximations of the (rotation) number α.

PROPOSITION 4.1.

(i) *If* $\nu(\alpha) < \infty$ *then* $\underline{\dim}_{(1/t)}(S^1) \leqslant \nu(\alpha) \leqslant \overline{\dim}_{(1/t)}(S^1)$.

(ii) *Furthermore, if the representation of* α *in the form of the continued fraction*

$$[a_1, a_2, \ldots, \alpha_i, \ldots]$$

has bounded elements, $0 < a_i \leqslant K_0 < \infty$, $i = 1, \ldots,$ *then*

$$\underline{\dim}_{(1/t)}(S^1) = \nu(\alpha) = \overline{\dim}_{(1/t)}(S^1).$$

PROOF. (i) Let $\varepsilon > 0$ and let n be the least positive integer number such that $\operatorname{dist}(x, f^n x) < \varepsilon$, $x \in S^1$. It means that $\tau(I_\varepsilon) = n$ where I_ε is the interval of the length 2ε centered at x. If we let $\nu = \nu(\alpha) - \delta$, with $\delta > 0$ an arbitrary small number, then the inequality (4.17) has an infinite sequence of solutions (m, n), with the positive value of n. Thus, let us consider the sequence $\varepsilon = \varepsilon_n = n^{-\nu}$, with n being such a solution for (4.17). Then, for $\xi(t) = 1/t$

$$R_\xi\big(0, q, \varepsilon_n, S^1\big) = \sum_{i=1}^{N(n)} \frac{1}{n^q} + \mathcal{O}\big(n^{-q}\big),$$

where $N(n) = \lfloor n^\nu/2 \rfloor$, thus

$$R_\xi\big(0, q, \varepsilon_n, S^1\big) = n^{-q+\nu(\alpha)-\delta}\Big(\frac{1}{2} + \mathcal{O}\big(n^{-2\nu}\big)\Big) + \mathcal{O}\big(n^{-q}\big).$$

Since

$$\underline{\lim}_{\varepsilon\to 0} R_\xi\big(0, q, \varepsilon, S^1\big) \leqslant \lim_{n\to\infty} R_\xi\big(0, q, \varepsilon_n, S^1\big) \leqslant \overline{\lim}_{\varepsilon\to 0} R_\xi\big(0, q, \varepsilon, S^1\big)$$

it follows that $\underline{\dim}_\xi(S^1) \leqslant \nu(\alpha) - \delta \leqslant \overline{\dim}_\xi(S^1)$ and, due to the arbitrariness of δ, statement (i) follows.

(ii) Let us order the solutions (m_k, n_k) of Eq. (4.17) in such a way that $n_{k+1} \geqslant n_k$, $k \in \mathbb{Z}_+$. The second assumption of the proposition implies that there exists a positive constant K such that $n_{k+1}/n_k \leqslant K < \infty$, $k \geqslant 1$. Let us consider now that for every sufficiently small $\varepsilon > 0$ there exists an integer $k = k(\varepsilon)$ such that $\varepsilon \in [n_{k+1}^{-\nu}, n_k^{-\nu}]$. Thus, $\tau(I_\varepsilon) \in [n_k, n_{k+1}]$ and

$$R_\xi\big(0, q, \varepsilon, S^1\big) \geqslant \sum_{i=1}^{N(n_k)} \frac{1}{n_{k+1}^q} \geqslant n_{k+1}^{-q+\nu(\alpha)-\delta}\Big(\frac{1}{2} + \mathcal{O}\big(n_{k+1}^{-2\nu}\big)\Big).$$

Since ε is arbitrary and $k \to \infty$ as $\varepsilon \to 0$, it follows that $\underline{\dim}_\xi(S^1) \geqslant \nu(\alpha)$. This, together with statement (i), allows us to conclude that $\underline{\dim}_\xi(S^1) = \nu(\alpha)$.

Set $\mu = \nu(\alpha) + \delta$ and let (m_k, n_k) be as in the previous paragraph. Then by definition of $\nu(\alpha)$ we have

$$|n_k\alpha - m_k| \geqslant n_k^{-\mu} \tag{4.19}$$

for every $k \geqslant k_0$, where k_0 is some positive number. Inequality (4.19) means that $\operatorname{dist}(x, f^{n_k}x) > n_k^{-\mu}$. For every $\varepsilon > 0$ there exists $k = k(\varepsilon)$ such that $\varepsilon \in [n_{k+1}^{-\mu}, n_k^{-\mu}]$. Then, by inequality (4.19) we have that $\tau(I_\varepsilon) > n_k$, where I_ε is the interval of length 2ε centered at point x. Therefore

$$\begin{aligned}
R_\xi(0, q, \varepsilon, S^1) &\leqslant \sum_{i=1}^{\lceil 1/(2\varepsilon)\rceil} \xi\big(\tau(I_\varepsilon)\big)^q \leqslant \sum_{i=1}^{\lceil n_{k+1}^{\mu}/2\rceil} n_k^{-q} \\
&\leqslant n_k^{-q} n_{k+1}^{\mu}\left(\frac{1}{2} + \mathcal{O}\big(n_{k+1}^{-2\mu}\big)\right) \\
&\leqslant K^{\mu} n_k^{-q+\nu(\alpha)+\delta}\left(\frac{1}{2} + \mathcal{O}\big(n_{k+1}^{-2\mu}\big)\right).
\end{aligned}$$

Since ε is arbitrary and $k \to \infty$ as $\varepsilon \to 0$, we get $\overline{\dim}_\xi(S^1) \leqslant \nu$ and, due to the arbitrariness of $\delta > 0$, $\overline{\dim}_\xi(S^1) \leqslant \nu(\alpha)$. □

The proposition tells us that $\langle\tau(u)\rangle \sim \varepsilon^{-1/\nu(\alpha)}$, $e \ll 1$, $\operatorname{diam} u = \varepsilon$, that is completely consistent with Eq. (4.6).

Consider now a case of the anomalously fast approximation of α by rational numbers. For example, assume that α is chosen in such a way that the inequality

$$\left|\alpha - \frac{m}{n}\right| < \frac{1}{n}e^{-bn} \tag{4.20}$$

holds for infinitely many relatively prime pairs (m, n), $n > 0$, provided that $b < b_0$, and

$$\left|\alpha - \frac{m}{n}\right| \geqslant \frac{1}{n}e^{-bn} \tag{4.21}$$

if $b > b_0$ and $n \geqslant n_0 > 0$.

PROPOSITION 4.2. *For the gauge function $\xi(t) = e^{-t}$,*

$$\underline{\dim}_\xi(S^1) \leqslant b_0 \leqslant \overline{\dim}_\xi(S^1).$$

PROOF. The scheme of the proof is the same as above. Set $\varepsilon_n = e^{-bn}$, $b = b_0 - \delta$, $\delta > 0$. Then

$$\tau\big(B(x,\varepsilon)\big) = n, \quad R_\xi\big(0, q, \varepsilon_n, S^1\big) = \sum_{i=1}^{N_2(n)} e^{-qn} + \mathcal{O}\big(e^{-BN}\big),$$

where $N_2(q) = \lfloor e^{bn}/2 \rfloor$, i.e.,

$$R_\xi\big(0, q, e^{-bn}, S^1\big) = e^{b-qn}\left(\frac{1}{2} + \mathcal{O}\big(e^{-2bn}\big)\right).$$

Therefore $\underline{\dim}_\xi(S^1) \leqslant b_0 - \delta \leqslant \overline{\dim}_\xi(S^1)$. □

As a corollary of the proposition we obtain that if $\underline{\dim}_\xi(S^1) = \overline{\dim}_\xi(S^1)$ then $\langle \tau(u) \rangle \sim -\log \varepsilon / b_0$, $\operatorname{diam} u = \varepsilon \ll 1$ (compare with Eq. (4.7)).

In the same way, we can show that if

$$\left| \alpha - \frac{m}{n} \right| \sim e^{-b_0 q^2},$$

then $\overline{\dim}_\xi(S^1) = b_0$, if the gauge function is $\xi(t) = e^{-t^2}$, i.e., $\langle \tau(u) \rangle \sim \sqrt{-\log \varepsilon / b_0}$, $\operatorname{diam} u = \varepsilon$, $\varepsilon \ll 1$, etc.

These examples show that the gauge functions $\xi(t)$ have to be chosen according to the rate of approximation of the rotation number α by rational numbers, otherwise the dimensions $\overline{\dim}_\xi(S^1) = \infty$ or 0. For example, if α satisfies Eq. (4.20) but $\xi(t) = 1/t$ then $\underline{\dim}_\xi(S^1) = \infty$. But, since Diophantine numbers (numbers satisfying Eq. (4.18) with $\nu(\alpha) < \infty$) form a set of full measure on any interval, then the function $1/t$ is presumably the most appropriate function for minimal sets on the circle [28].

4.6.3. Denjoy example

It consists of a C^1-smooth diffeomorphism, say f, of the circle $S^1 = \{x \pmod 1\}$ which generates a dynamical system with a non-trivial minimal Cantor-like set, say X (see, e.g., Refs. [73,90,48]). The idea of the construction is to start with a rigid rotation f_α on the circle, with the rotation number α, and replace points of one orbit by suitably chosen intervals. Let $\{x_n\}$ be an orbit, i.e., $x_n = f_\alpha^n(x_0)$, $n \in \mathbb{Z}$, and let L_n be the length of the nth interval, say I_n. Fix a value of γ, $0 < \gamma < 1$, and let $L_n = (|n| + C)^{-1/\gamma}$, where C is a constant which will be chosen below. According to the Denjoy construction (see Ref. [73]) we have to choose the constant C in such a way that

$$\sum_{n \in Z} L_n < 1 \tag{4.22}$$

(the length of S^1 equals 1). Note that

$$\begin{aligned}\sum_{n\in\mathbb{Z}} L_n &= C^{-1/\gamma} + 2\sum_{n=1}^{\infty}(N+C)^{-1/\gamma} \\ &\leqslant C^{-1/\gamma} + 2\int_0^{\infty}(x+C)^{-1/\gamma}\,dx \\ &= C^{-1/\gamma}\left(1+\frac{2\gamma}{1-\gamma}C\right) < 1\end{aligned}$$

if

$$C > C_*(\gamma), \tag{4.23}$$

where $C_*(\gamma)$ is the unique root of the equation

$$1+\frac{2\gamma}{1-\gamma}c = c^{1/\gamma}.$$

Let us remark that γ is the Hölder exponent of f' (see Ref. [73]).

Set $L = \sum_{n\in Z} L_n$. Now, we have to blow up the orbit x_n to the intervals I_n so that they are ordered in the same way as the points x_n and so that the distance between any two intervals I_m and I_n is exactly

$$(1-L)\,\mathrm{dist}(x_m, x_n) + \sum_{x_k\in(x_m,x_n)} L_k. \tag{4.24}$$

It can be seen from Eq. (4.24) that not only rotation number α and the number $\nu(\alpha)$, reflecting the rate of decreasing $\mathrm{dist}(x_m, x_n)$, but also the number γ, reflecting the rate of decreasing of the length L_n, has to be essential for Poincaré recurrences. Roughly speaking if $1/\gamma$ is large enough, we may neglect the dynamics inside the intervals I_n and treat them as points.

PROPOSITION 4.3. *If* $\nu(\alpha) < \infty$, $\xi(t) = 1/t$ *and* $\gamma < (\nu(\alpha)+1)^{-1}$ *then*

$$\nu(\alpha)\,\underline{\dim}_B(X) \leqslant \overline{\dim}_\xi(X) \leqslant \nu(\alpha)\,\overline{\dim}_B(X).$$

PROOF. Let $\bar{a} = \overline{\dim}_B X$, $\underline{a} = \underline{\dim}_B X$. Denote by $y_n' < y_n''$ the endpoints of the interval I_n and by y_n either y_n' or y_n''.

(1) Let $\mu = \nu(\alpha) + \delta$, $0 < \delta \ll 1$. Then the inequality (4.19) holds for any sufficiently large integer number n, i.e.,

$$\mathrm{dist}(x_n, x_m) \geqslant |m-n|^{-\mu}. \tag{4.25}$$

Choose $\varepsilon(\ell) = (1-L)/\ell^{\mu}$, $\ell \gg 1$, and consider a finite cover $G = \{u_i\}$ of X by open intervals u_i, $\operatorname{diam} u_i = \varepsilon$. Then, thanks to Eqs. (4.24) and (4.25), we have for any endpoints $y_m, y_n \in u_i$

$$(1-L)\frac{1}{\ell^{\mu}} = \varepsilon(\ell) > \operatorname{dist}(y_m, y_n) \geqslant (1-L)\operatorname{dist}(x_m, x_n) \\ \geqslant |m-n|^{-\mu}(1-L).$$

Thus, $\ell^{\mu} < |n-m|^{\mu}$. Therefore $\tau(u_i) > \ell$. Consequently,

$$R\big(q, \varepsilon(\ell), \xi\big) \leqslant \sum_i \frac{1}{\tau(u_i)^q} \leqslant \frac{1}{\ell^q} N(\varepsilon) \leqslant (\text{const})\,\frac{1}{\ell^q}\varepsilon(\ell)^{-(\bar{a}+\delta)} \\ \leqslant (\text{const})\,\ell^{-q+\mu(\bar{a}+\delta)}.$$

Hence, $R(q, \varepsilon, \xi) \ll 1$ if $q > \mu(\bar{a}+\delta)$ and due to arbitrariness of ℓ,

$$\overline{\dim}_{\xi} X \leqslant \big(\nu(\alpha)+\delta\big)(\bar{a}+\delta).$$

Therefore,

$$\overline{\dim}_{\xi} X \leqslant \nu(\alpha)\,\bar{a}.$$

(2) Let $\nu = \nu(\alpha) - \delta$. Then there is a sequence $(m_k, n_k)_{k \geqslant 1}$ of solutions to (4.17), such that

$$\operatorname{dist}(x_m, x_{m+n_k}) \leqslant n_k^{-\nu}$$

for any k and m. Furthermore, we may assume without loss of generality that the point x_{m+n_k} lies on the right of x_m for any k, m and the sequence x_{m+n_k} is monotone (see, e.g., Refs. [90] and [48]). Consider a finite cover $\{u_i\}$, $\operatorname{diam} u_i = \varepsilon$, where $\varepsilon = 2(1-L)/n_k^{\nu}$. Then for every u_i there exist an endpoint y_n, such that

$$\operatorname{dist}(y_n, a_i) < \frac{1}{2}(1-L)\frac{1}{n_k^{\nu}} \tag{4.26}$$

where a_i is the left endpoint of the interval u_i. Moreover, if the point $x_m \in (x_n, x_{n+n_k})$ then

$$|m| > K_n n_k,$$

where K_n is some constant. We have

$$\operatorname{dist}\big(y_n, f^{n_k} y_n\big) \leqslant (1-L)\operatorname{dist}(x_n, x_{n+n_k}) + \sum_{x_m \in (x_n, x_{n+n_k})} L_m \\ \leqslant (1-L)\frac{1}{n_k^{\nu}} + 2\int_{K_n n_k}^{\infty} \frac{dx}{(x+C)^{1/\gamma}} \\ \leqslant (1-L)\frac{1}{n_k^{\nu}} + K_2\left(\frac{1}{n_k}\right)^{1-1/\gamma},$$

where K_2 is a constant. Therefore, if $1 - 1/\gamma > \nu$, i.e. $\gamma < (1+\nu)^{-1}$, then

$$\operatorname{dist}\bigl(y_n, f^{n_k} y_n\bigr) < \frac{3}{2}(1-L)\frac{1}{n_k^{\nu}} < \varepsilon \tag{4.27}$$

provided that $k \gg 1$. The inequalities (4.27) and (4.26) imply that $f^{n_k} y_n \in u_i$ (as well as y_n), thus

$$\tau(u_i) \leqslant n_k, \tag{4.28}$$

and

$$\sum_i \frac{1}{\tau(u_i)^q} \geqslant N(\varepsilon)\frac{1}{n_k^q}.$$

Since $\{u_i\}$ is an arbitrary cover, then

$$\begin{aligned} R(q,\varepsilon,\xi) &\geqslant N(\varepsilon)\frac{1}{n_k^q} \geqslant (\text{const})\,\frac{1}{n_k^q}\varepsilon^{-(\underline{a}-\delta)} \\ &\geqslant (\text{const})\, n_k^{-q+\nu(\underline{a}-\delta)} \end{aligned}$$

and $R(q,\varepsilon,\xi) \gg 1$ if $q < \nu(\underline{a}-\delta)$. Therefore

$$\overline{\dim}_\xi X \geqslant \bigl(\nu(\alpha)+\delta\bigr)(\bar{a}-\delta). \qquad \square$$

An immediate consequence of Proposition 4.3 is the following.

COROLLARY 4.2. *Since* $\underline{\dim}_B(X) = \overline{\dim}_B(X) = \dim_B(X) = 1$, *then*

$$\overline{\dim}_\xi(X) = \nu(\alpha).$$

REMARK 4.1. We think that for linearly recurrent dynamical systems [46] the right gauge function is also $\xi(t) = 1/t$, but it not settled yet.

4.6.4. Multidimensional rotation

Consider the map

$$f : (x_1,\dots,x_n) \mapsto (x_1',\dots,x_n'), \quad x_k' = x_k + \alpha_k \ (\text{mod } 1),\ k = 1,\dots,n,$$

of the n-dimensional torus, S^n, where all numbers $\alpha_1,\dots,\alpha_n$ are irrational and linearly independent over the field of rational numbers. Thus, the torus is a minimal set for the map f and all orbits are recurrent.

Let

$$\|\ell\alpha_k\| = \inf_{m\in\mathbb{Z}} |\ell\alpha_k - m|, \quad k = 1,\dots,n,$$

be the distance from the nearest integer, and introduce a rate of approximation of the vector $\bar{\alpha} = (\alpha_1, \ldots, \alpha_n)$ by integer vectors: set

$$\nu(\bar{\alpha}) = \sup\Big\{\nu\colon \max_k \|\ell\alpha_k\| < \ell^{-\nu} \text{ has infinitely many integer solutions}\Big\}.$$

The following proposition can be proved in the same way as Proposition 4.17.

PROPOSITION 4.4. *If* $\nu(\bar{\alpha}) < \infty$ *then* $\overline{\dim}_{(1/t)}(S^n) = \nu(\bar{\alpha})$.

Thus, we can expect that

$$\big\langle \tau(u_i)^{-\ell\nu(\bar{\alpha})}\big\rangle \sim \varepsilon^{\ell},$$

i.e.,

$$\big\langle \tau(u_i)\big\rangle \sim \varepsilon^{-1/\nu(\bar{\alpha})}.$$

Here $\operatorname{diam} u_i = \varepsilon \ll 1$.

Let us remark that the vectors $\bar{\alpha}$ with $\nu(\bar{\alpha}) < \infty$ form a sufficiently large set. For example, it is known that its Hausdorff dimension is positive,

$$\dim_H J(\beta) = \frac{n+1}{\beta+1},$$

where $J(\beta) = \{\bar{\alpha}\colon \nu(\bar{\alpha}) = \beta\}$, see [69].

PART II

ZERO-DIMENSIONAL INVARIANT SETS

Chapter 5

Uniformly Hyperbolic Repellers

Let (F, f) be a dynamical systems over a Cantor set $F \subset \mathbb{R}^d$ whose geometric construction is described symbolically *à la* Moran with a function λ which is uniformly contracting, i.e., $\lambda_{\max} < 1$. Such sets appear as hyperbolic repellers, i.e., locally maximal invariants sets for which the stable set is just the repeller. In other words, the set is repelling in every direction in phase space. Let us further assume that f has the specification property and has positive topological entropy (for which we use $\xi(t) = e^{-t}$ in (4.3)).

To apply the formalism we have to make sure that the set function $\mathcal{M}_\xi(\alpha, q, \cdot)$ is an outer measure. Consider the distance defined by the property that the diameter of any n-cylinder is $\exp(-n)$. The set function $\mathcal{M}_\xi(\alpha, q, \cdot)$ is not an outer measure for the parameters range $q < 0$ and $\alpha \geqslant -q$. To see it, remark that such system (F, f) contains a subsystem $(E, f|E)$ which is not minimal and has arbitrarily small topological entropy, such that $\mathcal{M}_\xi(\alpha, q, E) < \infty$. Moreover, since $(E, f|E)$ is not minimal, then there exists an orbit $\{\underline{x}\} \subset E$ such that $\inf\{d(x_0, x_n)\colon n > 0\} > 0$. This implies that $\mathcal{M}_\xi(\alpha, q, \{\underline{x}\}) = \infty$. For $q < 0$ and $\alpha < -q$ the measure of a specified system is ∞: we always have n-cylinders with return time not greater than n. On the other hand it is an outer measure for $q > 0$ and all α. But for $\alpha < 0$ the measure is infinite if the set contains a periodic point. Thus, the only interesting parameter range to consider is $\alpha > 0$ and $q \geqslant 0$.

THEOREM 5.1. *Assume that set F either has the controlled-packing property or it satisfies the gap condition* (3.25)–(3.26). *Let the system (F, f) be topologically conjugate to a subshift (Ω, σ) with the specification property and positive topological entropy. Then, for $\xi(t) = e^{-t}$ and the parameter region $q \geqslant 0$ and $\alpha \geqslant 0$, the spectrum $\alpha_c(q)$ is the solution of the equation*

$$P_\Omega(\alpha \log \lambda) = q. \tag{5.1}$$

The dimension for Poincaré recurrences coincides with the topological entropy of the subshift (Ω, σ), i.e., $q_0 = h_{\text{top}}(\sigma|\Omega)$.

COROLLARY 5.1. *Formula* (5.1) *is valid for every fractal set F resulting from a one-dimensional Moran construction.*

It is a direct consequence of Lemma 3.2.

5.1. Connection with the multifractal spectrum of Lyapunov exponents

When (F, f) in Theorem 5.1 is a d-dimensional conformal repeller [97,15], there exists a relation between the entropy spectrum of Lyapunov exponents and the spectrum of dimensions for Poincaré recurrences.

Let us remind that a conformal repeller F is an f-invariant set such that:

(i) $Df(x) = L(x)\mathrm{Isom}(x)$, $x \in F$, where $L(x) > 1$ is a number and $\mathrm{Isom}(x)$ is an isometry, and
(ii) F is locally-maximal, i.e., there is a neighborhood $U \supset F$ such that $\bigcap_{i\geqslant 0} f^i(U) = F$.

Let $x \in F$ and denote by $\Lambda(x)$ the expression

$$\lim_{n\to\infty} \frac{1}{n} \sum_{j=0}^{n-1} \frac{1}{d} \log \|D_{f^j(x)} f\|,$$

whenever this limit exists, and call it the Lyapunov exponent. We denote by E the set of points $x \in F$ where the Lyapunov exponent does not exist. For $\beta \in \mathbb{R}$ we set

$$E_\beta := \{x \in F \setminus E\colon\ \Lambda(x) = \beta\}.$$

We have that $F = E \cup \bigcup_{\beta\in\mathbb{R}} E_\beta$.

Let us define the *entropy spectrum of the Lyapunov exponents* by $\eta(\beta) := h_{\mathrm{top}}(\sigma|E_\beta)$. Here, h_{top} denotes the topological entropy for non-compact sets as it is defined in the second example of Section 4.1.1 [97,29]. Following the proofs of Theorem 1 and Theorem 2 in [27] one can show that in this case

$$\eta(\beta) = \inf_\alpha \{P_F(\alpha\phi) + \alpha\beta\}$$

where $\phi(x) = -(1/d)\log\|D_x f\|$. In view of Theorem 5.1 $\alpha_c(q)$ fulfills the equation $q = P_\Omega(\alpha_c \log(\lambda(\omega)))$. Since

$$P_\Omega(\alpha_c \log(\lambda(\omega))) = P_F\left(-\frac{\alpha_c}{d} \log\|D_x f\|\right)$$

we get

$$\eta(\beta) = \min_\alpha \{\alpha_c^{-1}(\alpha) + \alpha\beta\}.$$

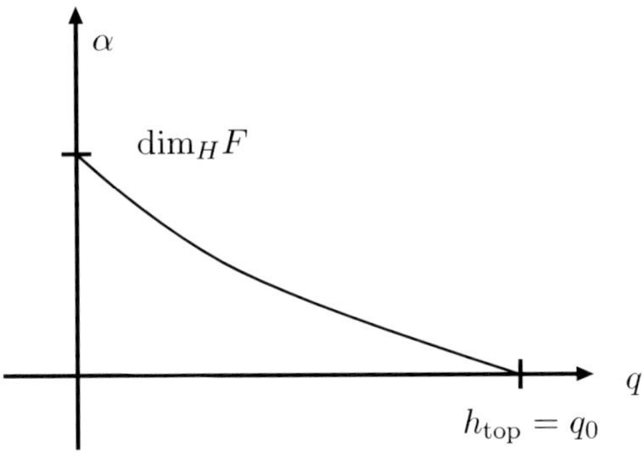

Figure 5.1.

The spectrum $\eta(\beta)$ is strictly concave and defined on a closed interval. Hence, the following result holds.

THEOREM 5.2. (See [2].) *The entropy spectrum of Lyapunov exponents $\eta(\beta)$ and the inverse α_c^{-1} of the spectrum for Poincaré recurrences form a Legendre-transform pair.*

In this case the spectrum α_c is strictly decreasing. It is strictly convex iff $\phi(x) = -(1/d)\log \|D_x f\|$ is not cohomologous to a constant. In the latter case the support of $\eta(\beta)$ reduces to a point and $\alpha_c(q)$ is linear:

$$\alpha_c(q) = \frac{q - h_{\text{top}}(\sigma\,|\,\Omega)}{\log\lambda}.$$

For all the systems we consider we have that $\alpha_c(h_{\text{top}}(f\,|\,F)) = 0$ and $\alpha_c(0) = \dim_H(F)$. Hence, for the quasi-conformal repellers considered in this paragraph and for $0 \leqslant q \leqslant h_{\text{top}}(\sigma\,|\,\Omega)$ the critical value $\alpha_c(q)$ lies in the interval $[0, \dim_H(F)]$. Let us emphasize that if $q = 0$ then $\alpha_c(0) = -h_{\text{top}}/\log\lambda = \dim_H \Omega_A$, thus, the spectrum of dimensions can be treated as a family "joining" the extreme values: the topological entropy and the Hausdorff dimension (see Figure 5.1).

5.2. Spectra under the controlled-packing condition

The next two lemmas imply that, given $q \geqslant 0$, $\alpha_c(q)$ satisfies the equation $q = P_S(\alpha_c(q)\log(\lambda))$ provided $\alpha_c(q) \geqslant 0$. This is the statement of Theorem 5.1.

LEMMA 5.1. *For a fixed $q \geqslant 0$ the quantity $\mathcal{M}(\alpha, q) = 0$ for every $\alpha \geqslant 0$ such that $P_S(\alpha\log(\lambda)) < q$.*

LEMMA 5.2. *For a fixed* $q \geqslant 0$ *the quantity* $\mathcal{M}(\alpha, q) = \infty$ *for every* $\alpha \geqslant 0$ *such that* $P_S(\alpha \log(\lambda)) > q$.

As we will see below, it is easy to prove Lemma 5.1 while Lemma 5.2 demands some extra work. The concept behind the proofs of each one of these lemmas is the use of properties of the topological pressure.

5.2.1. Proof of Lemma 5.1

Given $\varepsilon > 0$ let $\mathcal{C} \in \mathcal{B}_\varepsilon$ be a cover of F by balls such that for any $B \in \mathcal{C}$, $B \cap F = \chi[\underline{\omega}]$ and $|\underline{\omega}| =: n_\varepsilon$ (independent of B). Trivially $n_\varepsilon \to \infty$ as $\varepsilon \to 0$. Because of (3.6),

$$\mathcal{M}(\alpha, q, \varepsilon) \leqslant \bar{c} \sum_{\substack{[\underline{\omega}] \in S \\ |\underline{\omega}| = n_\varepsilon}} \exp\left(-q\tau\big(\chi([\underline{\omega}])\big) + \alpha \sum_{j=0}^{n_\varepsilon - 1} \log \lambda\big(\sigma^j \omega\big) \right). \tag{5.2}$$

Here and in the following for each cylinder $[\underline{\omega}] \subset S$ we choose an arbitrary $\omega \in [\underline{\omega}]$. Because of the specification property, the cylinder $[\underline{\omega}]$ has first return time k if it contains a periodic sequence $\omega = \omega_0, \ldots, \omega_{k-1}, \omega_0 \ldots$ of period k, and any other periodic sequences in it has a greater period. Denoting by $P_{n_\varepsilon, k}$ the set of cylinders of length n_ε having first return time k, we conclude that for $\alpha \geqslant 0$ and $q \in \mathbb{R}$

$$\begin{aligned} \mathcal{M}(\alpha, q, \varepsilon) &\leqslant \bar{c} \sum_{k=1}^{n_\varepsilon + n_0} \left(\exp(-qk) \sum_{[\underline{\omega}] \in P_{n_\varepsilon, k}} \exp\left(\alpha \sum_{j=0}^{n_\varepsilon - 1} \log \lambda\big(\sigma^j \omega\big) \right) \right) \\ &\leqslant \bar{c}_1 \sum_{k=1}^{n_\varepsilon + n_0} \sum_{[\underline{\omega}] \in P_{n_\varepsilon, k}} \exp\left(-qk + \alpha \sum_{j=0}^{k-1} \log \lambda\big(\sigma^j \omega\big) \right) \\ &\leqslant \bar{c}_1 \sum_{k=1}^{\infty} \sum_{\substack{[\underline{\omega}] \in S \\ |\underline{\omega}| = k}} \exp\left(-qk + \alpha \sum_{j=0}^{k-1} \log \lambda\big(\sigma^j \omega\big) \right), \end{aligned}$$

where the sums in the series are defined in (2.9) and $\bar{c}_1 = \bar{c}(\lambda_{\min})^{-\alpha n_0}$. Then

$$\mathcal{M}(\alpha, q, \varepsilon) \leqslant \bar{c}_1 \sum_{k=1}^{\infty} Z_k(-q + \alpha \log \lambda, S). \tag{5.3}$$

According to (2.10) and (2.11) for $q > P_S(\alpha \log \lambda)$ the expressions

$$Z_k(-q + \alpha \log \lambda, S)$$

converge exponentially to zero as $k \to \infty$. Hence, the series in the right hand side of (5.3) converges, and the result follows. □

Before proceeding with the proof of Lemma 5.2 we need the following technical result (let us remind that the length of a cylinder $[\underline{\omega}]$ is denoted by $|\underline{\omega}|$).

LEMMA 5.3. *Let $F \subset \mathbb{R}^d$ have controlled packing of cylinders with exponent a. Let G be a finite or countable cover of F by open balls. Then for $\alpha \geqslant 0$, $q \geqslant 0$ there is a positive constant C such that*

$$\mathcal{M}(\alpha, q, G) \geqslant C \sum_{[\underline{\omega}] \in \mathrm{CMax}(G)} \exp\bigl(-q|\underline{\omega}|\bigr)|\underline{\omega}|^{-a} \prod_{j=0}^{|\underline{\omega}|-1} \lambda\bigl(\sigma^j \omega\bigr)^{\alpha}. \tag{5.4}$$

PROOF. Let $B \in \mathcal{C}$. For an arbitrary $0 < \rho < 1$ we denote by N_B the integer for which the diameter

$$D := \max\bigl\{\bigl|\chi([\underline{\omega}])\bigr|\colon [\underline{\omega}] \in \mathrm{CMax}(B)\bigr\}$$

belongs to the interval $(\rho^{N_B+1}, \rho^{N_B}]$. Then condition (3.27) implies that

$$\begin{aligned}\sum_{[\underline{\omega}] \in \mathrm{CMax}(B)} \bigl|\chi([\underline{\omega}])\bigr|^{\alpha} &\leqslant C_0 \sum_{k=N_B}^{\infty} \rho^{k\alpha} k^a \\ &\leqslant C_0 \Bigl(\frac{D}{\rho}\Bigr)^{\alpha} \sum_{k=0}^{\infty} \bigl(\rho^{\alpha}\bigr)^k (N_B + k)^a \\ &\leqslant C_1 |B|^{\alpha} N_B^a,\end{aligned} \tag{5.5}$$

where

$$C_1 = \bigl(C_0/\rho^{\alpha}\bigr) \sum_{k=0}^{\infty} \rho^{k\alpha} (1+k)^a.$$

Taking into account that $|\chi([\underline{\omega}])| < \rho^{N_B}$ for every $[\underline{\omega}] \in \mathrm{CMax}(B)$, we obtain from (5.5) that

$$\begin{aligned}|B|^{\alpha} &\geqslant C_1^{-1} N_B^{-a} \sum_{[\underline{\omega}] \in \mathrm{CMax}(B)} \bigl|\chi([\underline{\omega}])\bigr|^{\alpha}, \\ &\geqslant C_2 \sum_{[\underline{\omega}] \in \mathrm{CMax}(B)} \frac{|\chi([\underline{\omega}])|^{\alpha}}{|\log|\chi([\underline{\omega}])||^a},\end{aligned} \tag{5.6}$$

where $C_2 = C_1^{-1} |\log \rho|^a$.

In view of the specification property $\tau(B) \leqslant \tau([\underline{\omega}]) \leqslant |\underline{\omega}| + n_0$ for all $[\underline{\omega}] \in \mathrm{CMax}(B)$. Inequality (3.6) implies that

$$\bigl|\log\bigl|\chi([\underline{\omega}])\bigr|\bigr| \leqslant |\log \underline{d}| + |\underline{\omega}|\,|\log \lambda_{\min}|,$$

where $\lambda_{\min} := \min_{\omega\in S}\{\lambda(\omega)\}$. Therefore,

$$|B|^{\alpha}\exp\bigl(-q\tau(B)\bigr) \geqslant C_3 \sum_{[\underline{\omega}]\in\mathrm{CMax}(B)} \exp\bigl(-q|\underline{\omega}|\bigr)\bigl|\chi([\underline{\omega}])\bigr|^{\alpha}|\underline{\omega}|^{-a},$$

where

$$C_3 = \frac{C_2}{(|\log \lambda_{\min}| + |\log \underline{d}|)^{a}}.$$

Finally, since $q \geqslant 0$

$$\begin{aligned}\sum_{B\in G}\sum_{[\underline{\omega}]\in\mathrm{CMax}(B)} &\exp\bigl(-q|\underline{\omega}|\bigr)\bigl|\chi([\underline{\omega}])\bigr|^{\alpha}|\underline{\omega}|^{-a}\\ &\geqslant \underline{d}^{\alpha}\sum_{[\underline{\omega}]\in\mathrm{CMax}(G)}\exp\bigl(-q|\underline{\omega}|\bigr)|\underline{\omega}|^{-a}\prod_{j=0}^{|\underline{\omega}|-1}\lambda\bigl(\sigma^{j}\omega\bigr)^{\alpha}\end{aligned}$$

and the result follows with $C = C_3\,\underline{d}^{\alpha}$. □

5.2.2. *Proof of Lemma 5.2*

We fix ε small enough to ensure that inequality (3.27) holds. Let G be a cover of F by open balls of radius less than or equal to ε. Thus, by Lemma 5.3

$$\begin{aligned}\mathcal{M}(\alpha,q,G) \geqslant C \sum_{[\underline{\omega}]\in\mathrm{CMax}(G)} \exp\Biggl(&-|\underline{\omega}|\bigl(q + a\log|\underline{\omega}|/|\underline{\omega}|\bigr)\\ &+\alpha\sum_{j=0}^{|\underline{\omega}|-1}\log\lambda\bigl(\sigma^{j}\omega\bigr)\Biggr),\end{aligned} \tag{5.7}$$

where $\mathrm{CMax}(G)$ is the cover of S by all B-maximal cylinders $B \in G$. Let us remark that $|\chi([\underline{\omega}])| \leqslant \varepsilon$ for any cylinder $[\underline{\omega}] \in \mathrm{CMax}(G)$. Let

$$n_{\varepsilon} = \min\bigl\{|\underline{\omega}|\colon\ [\underline{\omega}] \in \mathrm{CMax}(G)\bigr\}.$$

Since

$$a\log|\underline{\omega}|/|\underline{\omega}| \leqslant a\log n_{\varepsilon}/n_{\varepsilon}$$

we have that for all $[\underline{\omega}] \in \mathrm{CMax}(G)$

$$\begin{aligned}\mathcal{M}(\alpha, q, G) &\geqslant C \sum_{[\underline{\omega}]\in \mathrm{CMax}(G)} \exp\Biggl(-|\underline{\omega}|\bigl(q + a\log(n_\varepsilon)/n_\varepsilon\bigr) \\ &\qquad + \alpha \sum_{j=0}^{|\underline{\omega}|-1} \log\lambda\bigl(\sigma^j\omega\bigr)\Biggr) \\ &= C\mathcal{Z}\bigl(q + a\log n_\varepsilon/n_\varepsilon, \phi, \mathrm{CMax}(G), S\bigr),\end{aligned} \tag{5.8}$$

where $\mathcal{Z}$ is defined in (2.16) and $\phi(\omega) = \alpha\log\lambda(\omega)$. Since $n_\varepsilon \to \infty$ as $\varepsilon \to 0$ we see that for all $\delta > 0$

$$\lim_{\varepsilon\to 0}\inf_{G\in\mathcal{B}_\varepsilon} \mathcal{M}(\alpha, q, C) \geqslant C \lim_{n\to\infty}\inf_{G\in G_n} \mathcal{Z}\bigl(q + \delta, \phi, \mathrm{CMax}(G), S\bigr), \tag{5.9}$$

where G_n is the class of all covers of S by cylinders with lengths greater than n. Hence, for all $\delta > 0$ the inequality $q + \delta < P_S(\alpha\log\lambda)$ implies

$$\lim_{\varepsilon\to 0}\inf_{G\in\mathcal{B}_\varepsilon} \mathcal{M}(\alpha, q, G) = \infty. \tag{5.10}$$

Theorem 5.1 under the controlled-packing condition follows from the two previous lemmas. □

5.3. Spectra under the gap condition

The strategy is to prove that for $d = 1$ all fractal sets resulting from a Moran construction satisfy the gap condition. Then, we prove that every fractal set F resulting from a strong Moran construction is equivalent to a fractal set in $\mathbb{R}^1$.

LEMMA 5.4. *Let (F, f) be a system topologically conjugate to a subshift (S, σ) and $F \subset \mathbb{R}^d$ a fractal set satisfying the gap condition. Then there exists a one-dimensional conformal repeller with the same spectrum of dimensions for Poincaré recurrences as (F, f).*

PROOF. We define a one-dimensional expanding map (J_S, g) as follows. Let I_i, $i = 0, \ldots, p-1$, be a collection of pairwise disjoint closed intervals in $\mathbb{R}$. Consider a piecewise expanding map $g : \bigcup_{i=0}^{p-1} I_i \to \mathbb{R}$ with branches $g_i : I_i \to \mathbb{R}$ such that for each i, $g(I_i) \supset \bigcup_{j=0}^{p-1} I_j$. The system $(\bigcup_{i=0}^{p-1} I_i, g)$ has a conformal repeller

$$J = \bigcap_{n=1}^{\infty} \bigcup_{\omega\in\Omega_p} g_{\omega_0}^{-1} \circ g_{\omega_1}^{-1} \circ \cdots \circ g_{\omega_{n-1}}^{-1}\bigl([0, L]\bigr),$$

which results from a Moran-like construction satisfying the conditions (M1)–(M3). The coding function $\chi' : S \to J$ is a topological conjugacy of the full shift (Ω_p, σ) with (J, g). Such a repeller exists for any mapping g with expanding branches g_i. In particular, we choose g such that

$$\lambda(\omega) = \frac{1}{|g_i'(\chi'(\omega))|} < 1$$

for every $\omega \in S \subset \Omega_p$, where λ and S correspond to the construction of F. We also assume that for every word $\underline{\omega} = (\omega_0, \ldots, \omega_{i-1})$ the Δ-sets

$$\Delta_{\omega_0 \ldots \omega_{i-1}} := g_{\omega_0}^{-1} \circ \cdots \circ g_{\omega_{i-1}}^{-1}\big([0, L]\big)$$

satisfy the extra conditions (3.25) and (3.26) of a strong-Moran construction, namely

$$\operatorname{dist}(\Delta_{\underline{i}\,\omega_i}, \Delta_{\underline{i}\,\omega_i'}) \geqslant G \operatorname{diam} \Delta(w_0, \ldots, w_{i-1}), \tag{5.11}$$

where dist is the usual absolute value metric in $\mathbb{R}$. The resulting one-dimensional system is denoted by $(J_S, g|J_S)$, where $J_S = \chi'(S)$.

The coding function corresponding to the construction of F is $\chi : S \to F$. By construction, the systems (F, f) and $(J_S, g|J_S)$ are topologically conjugate and the conjugacy is given by

$$\chi' \circ \chi^{-1} =: \kappa : F \to J_S.$$

Since the rates of contraction in the Moran constructions for F and J_S are the same, the systems (F, f) and $(J_S, g|J_S)$ have the same spectra of Poincaré recurrences. We consider the pushed-forward metric η on J_S

$$\eta(x, y) = \operatorname{dist}_d\big(\kappa^{-1}(x), \kappa^{-1}(y)\big), \tag{5.12}$$

where dist_d is the Euclidean metric on F. The conjugacy κ is by definition an isometry between the set F with the Euclidean metric and the set J_S with the metric η. This proves that (F, f) and $(J_S, g|J_S)$ with the pushed-forward metric η have the same spectra. Finally, the equivalence of the usual metric dist and the metric η on J_S is implied by inequality (5.11) and the fact that any two points x and x' in $J_S \subset \mathbb{R}$, satisfy

$$G \prod_{i=0}^{n-1} \lambda\big(\sigma^i \omega\big) \leqslant \operatorname{dist}(x, x') \leqslant \bar{c} \prod_{i=0}^{n-1} \lambda\big(\sigma^i \omega\big). \tag{5.13}$$

□

REMARK 5.1. Since the two systems are topologically conjugated the proof of Lemma 5.4 is independent of the function $\xi(t)$ in the definition of the outer measure $\mathcal{M}_\xi(\alpha, q, \cdot)$.

REMARK 5.2. We believe that the developed here machinery can be used to study Poincaré recurrences for, not only repellers, but other hyperbolic invariant sets of dynamical systems. In this direction the ideas and results of [20] will be very essential.

Chapter 6

Non-Uniformly Hyperbolic Repellers

If a repeller is hyperbolic then it can be obtained as the result of a geometric construction with rates of contraction uniformly smaller than 1. In this situations the Hausdorff dimension of the invariant set is the root of the Bowen's equation, and it was shown in Theorem 5.1 that the spectrum of dimensions for Poincaré recurrences satisfies the non-homogeneous Bowen's equation. Such a representation provides us a new possibility to study multifractal features of invariant sets. Moreover, Theorem 5.2 states that in the case of conformal repellers the entropy spectrum of Lyapunov exponents and the inverse function for the spectrum of dimensions for Poincaré recurrences form a Legendre-transform pair.

But if a repeller is non-hyperbolic or non-uniformly hyperbolic then there can be a point in it that belongs to an infinite sequence of balls accumulating to this point "with sub-exponential speed", and the corresponding rates of contraction are not disjoint from 1. In this Section we extend the previous results valid for hyperbolic systems to the case where the inductive geometric procedure is still governed by a specified subshift but rates of contraction contains 1 as a point of accumulation. Such a kind of problems was proposed in [8].

Usually, for non-hyperbolic and non-uniformly-hyperbolic systems there is no finite Markov partition, and we cannot describe our system in terms of topological Markov chains (TMC) with finitely many states. The problem thus arises to calculate the topological pressure, and so, the Hausdorff dimension and dimensions for Poincaré recurrences in those cases.

The goal in the present chapter is to extend the validity of Theorem 5.1 to the case where $\lambda_{\max} = 1$. So, let $\Lambda_c := \{\omega \in \Omega \mid \lambda(\omega) = 1\}$ and $F_c := \chi(\Lambda_c) \subset F$ be called critical sets. They are closed sets which we assume non-empty. There are two logical possibilities: either F_c contains or does not contain an orbit.

The validity of Theorem 5.1 for the case when the critical set F_c does not contain an orbit is proved in the next section. The extension is a direct consequence of the fact that cylinder sets form, in this case, a basis. When the critical set F_c does contain orbits we cannot use Lemma 6.1 below. Nevertheless, we identify situations where one can approximate an invariant set by a sequence of topological Markov chains. Since it is well known how to calculate the topological pressure for TMCs, the approximation scheme allows one to calculate or/and estimate the topological pressure of the original system. This is done in Section 6.2.

6.1. The critical set does not contain an orbit

The extension of Theorem 5.1 follows almost straightforwardly. We only need to ensure that covers by cylinder sets form a basis. This follows from the next result.

LEMMA 6.1. *Assume that Λ_c does not contain an orbit. Then there exist constants $0 < \mu < 1$ and $C \geqslant 1$, and a positive integer number N_c, such that*

$$\max\left\{\prod_{j=0}^{n} \lambda(\sigma^j \omega) \colon \omega \in S\right\} \leqslant C\mu^n \quad \textit{for } n > N_c.$$

PROOF. Since Λ_c does not contain an orbit, for each $\omega \in \Omega$ there exists a minimal $n_\omega \in \mathbb{N}$ such that $\sigma^{n_\omega}(\omega) \notin \Lambda_c$. For each $k \in \mathbb{N}$ let

$$\Lambda_c^{(k)} \equiv \{\omega \in \Lambda_c \colon n_\omega = k\}.$$

We may write Λ_c as the disjoint union $\Lambda_c = \bigcup_{k=1}^{\infty} \Lambda_c^{(k)}$. Note that

$$\bigcap_{k=0}^{n} \sigma^k(\Lambda_c) = \bigcup_{k=n+1}^{\infty} \Lambda_c^{(k)}$$

is closed, therefore its complement in Λ_c is relatively open. Thus,

$$\left\{\bigcup_{k=1}^{n} \Lambda_c^{(k)} \colon n \in \mathbb{N}\right\}$$

is an open cover for the compact set Λ_c, implying that $\Lambda_c = \bigcup_{n=1}^{N_c} \Lambda_c^{(n)}$ for some $N_c \in \mathbb{N}$.

Now, being Λ_c and $\sigma(\Lambda_c^{(1)})$ disjoint compact sets, and σ a continuous function, for each $\varepsilon > 0$ sufficiently small there exists $\delta = \delta(\varepsilon) > 0$ such that

$$d\big(\omega, \Lambda_c^{(1)}\big) < \delta \quad \Rightarrow \quad d\big(\sigma(\omega), \Lambda_c\big) \geqslant \varepsilon.$$

For $\varepsilon > 0$ sufficiently small, define $\lambda_1 \equiv \max\{\lambda(\omega) \colon d(\omega, \Lambda_c) \geqslant \varepsilon\}$, which is smaller than one. For each $k \in \{0, 1, \ldots, N_c - 1\}$ let

$$U_k \equiv \sigma^{-k}\big\{\omega \in \Omega \colon d\big(\omega, \Lambda_c^{(1)}\big) < \delta(\varepsilon)\big\},$$

and $U \equiv \bigcup_{k=0}^{N_c-1} U_k$. This is an open set containing Λ_c, then

$$\lambda_2 \equiv \max\{\lambda(\omega)\colon\ \omega \notin U\}$$

is smaller than one.

Finally, for each $\omega \in \Omega$ and every $n \geqslant N_c + 1$ we have

$$\prod_{j=0}^{n} \lambda(\sigma^j \omega) \leqslant \lambda_m^{\frac{n}{N_c+1}},$$

by taking $\lambda_m = \max\{\lambda_1, \lambda_2\}$, and the lemma follows with $\mu = \lambda_m^{\frac{1}{N_c+1}}$ and $C = \mu^{-1}$. □

COROLLARY 6.1. *For each $\varepsilon > 0$, there exists $n_\varepsilon \in \mathbb{N}$ such that* diam $(\chi[\underline{\omega}]) < \varepsilon$ *for all $\underline{\omega}$ such that $|\underline{\omega}| \geqslant n_\varepsilon$.*

Thus, if the critical set does not contain an orbit, then the cylinder sets form a basis. This result makes valid Lemma 3.1 and Theorem 5.1 to the present case (word by word, proofs are the same).

EXAMPLE 6.1. The mapping is inspired in the Manneville–Pomeau example (see [117]) and its critical set does not contain an orbit.

For $\delta \in (0, 1/2)$ define $I_0 = [0, 1/2 - \delta]$ and $I_1 = [1/2 + \delta, 1]$. Let a map $f : I_0 \cup I_1 \to [0, 1]$ be such that

(E1) f is strictly increasing and twice continuously differentiable in the interior of I_0 and I_1, and it is such that $f(I_0) = f(I_1) = [0, 1]$.

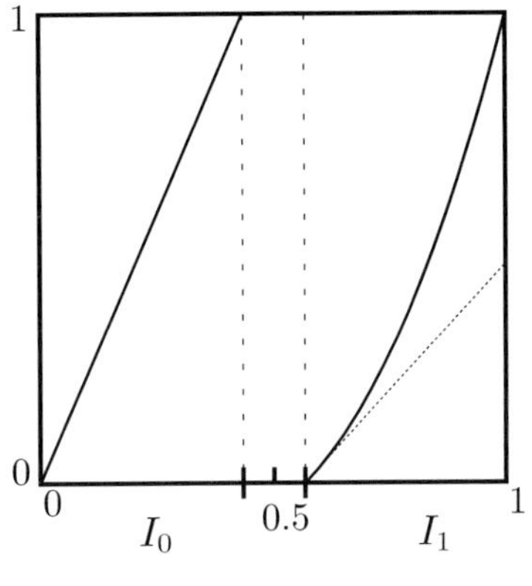

Figure 6.1.

(E2) $f'(x) > 1$ for all $x \in I_0 \cup (I_1 \setminus \{1/2+\delta\})$, $f'(1/2+\delta) = 1$ and $f''(x) \geqslant 0$ for all $x \in I_0 \cup I_1$.

See, for instance, the sketch in Figure 6.1. The repeller

$$F := \{x \in I_0 \cup I_1\colon\ f^n(x) \in I_0 \cup I_1\ \forall\, n \in \mathbb{N}\}$$

associated to the map is a Cantor set that can be obtained by a Moran construction, modeled by the full shift in two symbols as follows. The inverse functions of the two branches of f, $g_0 := (f|_{I_0})^{-1}$ and $g_1 := (f|_{I_1})^{-1}$, are contractions. Then, for each $\omega \in \mathbb{Z}_2^{\mathbb{N}}$ and $n \in \mathbb{N}$, let us consider

$$\Delta(\omega_0, \omega_1, \ldots, \omega_n) := g_{\omega_0} \circ g_{\omega_1} \circ \cdots \circ g_{\omega_n}([0,1])$$

to be the basic sets of a Moran construction for F with conjugacy $\chi : zz_2^{\mathbb{N}} \to F$,

$$\chi(\omega) = \bigcap_{n=0}^{\infty} g_{\omega_0} \circ \cdots \circ g_{\omega_n}([0,1]).$$

By Taylor's theorem, and applying the chain rule, there exists $\omega \in [(\omega_0, \omega_1, \ldots, w_n)]$ such that

$$\operatorname{diam} \Delta(\omega_0, \omega_1, \ldots, \omega_n) = \prod_{j=0}^{n} \lambda(\sigma^j \omega),$$

where $\lambda(\omega) := 1/|f'(\chi(\omega))|$.

It is easily verified that Λ_c is the singleton $\{1000000\ldots\}$, which is not a fixed point. So, Λ_c does not contain an orbit and conditions of Lemma 6.1 and Theorem 5.1 are satisfied. Thus,

$$P_{\mathbb{Z}_2^{\mathbb{N}}}(\alpha_c \log \lambda) = q. \qquad \square$$

6.2. The critical set contains an orbit

In this case Lemma 6.1 is not valid and Theorem 5.1 cannot be extended so straightforwardly as above. The alternative is to compute the spectra for Poincaré recurrences for a sequence of subsets $F_n \subset F$, modeled by subshifts $\Omega_n \subset \Omega$ that do not contain any critical point. Then, under reasonable assumptions, we will prove that $\alpha_c(q, F_n) \to \alpha_c(q, F)$ as $n \to \infty$.

The approximating sets, $F_n \subset F$ are defined as follows. For each $n \in \mathbb{N}$ let

$$G_n := \{\underline{\omega} \text{ is } \Omega\text{-admissible: } |\underline{\omega}| = n,\ [\underline{\omega}] \cap \Lambda_c = \emptyset\}.$$

Let Ω_n be a specified subshift $\Omega_n \subset \Omega$, such that $(\omega_0, \omega_1, \ldots, \omega_{n-1}) \in G_n$ for all $\omega \in \Omega_n$. It is easy to verify that $\Omega_n \subset \Omega_{n+1}$ for all $n \in \mathbb{N}$. We will also consider, for each $n \in \mathbb{N}$, the sets

$$\widetilde{G}_{n+1} := \{(\omega_0, \omega_1, \dots, \omega_n) \in G_{n+1}\colon (\omega_0, \omega_1, \dots, \omega_{n-1}) \notin G_n\}.$$

For each $n \in \mathbb{N}$, the *nth level approximation* to (F, f) is the sub-system (F_n, f), with $F_n := \chi(\Omega_n)$.

For each $n \in \mathbb{N}$ let

$$\lambda^{(n)} := \max\{\lambda(\omega)^n\colon \omega \in \Omega_n\},$$
$$\delta_n := \max\{k \leqslant n\colon \tau([\underline{\underline{\omega}}]) \geqslant n - k \ \forall \underline{\underline{\omega}} \in \widetilde{G}_n\}.$$

The quantity δ_n measures the delay in the Poincaré recurrence for cylinders of the nth level approximation. The sequence $\{(F_n, f)\}_{n\in\mathbb{N}}$ is *a good approximation* if the following conditions hold.

(H1) The critical set F_c has zero Hausdorff dimension.

(H2) For each $\omega \notin \Lambda_c$ there exists $n \geqslant \mathbb{N}$ such that $(\omega_0, \omega_1, \dots, \omega_{n-1})$ is Ω_n-admissible.

(H3) For each $q, \eta > 0$,

$$\sum_n \exp(q\delta_n + \eta \log \lambda^{(n)}) < \infty.$$

Condition (H3) establishes a relation between the speed of convergence of $\max_{\Omega_n} \lambda(\omega)$ and the Poincaré recurrence time of set $\Omega_n \setminus \Omega_{n-1}$.

THEOREM 6.1. *Assume that set F either has the controlled-packing property or it satisfies the gap condition* (3.25)–(3.26). *Let the system (F, f) be topologically conjugate to a subshift (Ω, σ) with the specification property and positive topological entropy. Let $\{(F_n, f)\}_{n\in\mathbb{N}}$ be a good approximation for (F, f). Then, for $\xi(t) = e^{-t}$,*

$$\alpha_c(q, \xi, F) = \sup_n \alpha_c(q, \xi, F_n)$$

in the parameter region $q \geqslant 0$ and $\alpha \geqslant 0$.

PROOF. Since $F \supset F_n$ for each $n \in \mathbb{N}$, we only have to prove that

$$\alpha_c(q, \xi, F) \leqslant \sup_n \alpha_c(q, \xi, F_n)$$

for each $q \geqslant 0$. Note that $q \mapsto \alpha_c(q, \xi, A)$ is a non-increasing function and, because of (H1), $\alpha_c(q, \xi, F_c) = 0$ for all $q > 0$. Therefore,

$$\alpha_c(q, \xi, F) = \alpha_c(q, \xi, F \setminus F_c)$$

for all $q \geqslant 0$.

For each $m \in \mathbb{N}$ let

$$G_m := \left\{ B(\underline{\omega}) \colon \underline{\omega} \in G_m \cup \bigcup_{n>m} \widetilde{G}_n \right\},$$

where $B(\underline{\omega})$ is an open ball such that $B(\underline{\omega}) \cap F = \chi([\underline{\omega}])$. Since $\chi : S \to F$ is a homeomorphism, hypothesis (M3) allows us to choose such open ball for each Ω-admissible word. Hypothesis (H2) ensures that G_m is a cover for $F \setminus F_c$, while (H3) and (M2) imply

$$\begin{aligned}\operatorname{diam}(G_m) &\leqslant \bar{c} \max \left\{ \prod_{j=0}^{n-1} \lambda(\sigma^j \omega) \colon \omega \in S_n,\ n \geqslant m \right\} \\ &\leqslant \bar{c} \max_{n \geqslant m} \lambda^{(n)} \to 0 \quad \text{as } m \to \infty.\end{aligned}$$

Then, for each $m \in \mathbb{N}$ we have

$$\begin{aligned}\mathcal{M}_{\xi, G_m}(\alpha, q, F \setminus F_c) &\equiv \sum_{B \in G_m} \exp\bigl(-q\tau(B)\bigr) \operatorname{diam}(B)^\alpha \\ &\leqslant \bar{c} \sum_{\underline{\omega} \in G_m} \exp\left(-q\tau\bigl(B(\underline{\omega})\bigr) + \alpha \sum_{j=1}^{m-1} \log \lambda\bigl(\sigma^j \omega\bigr) \right) \\ &\quad + \bar{c} \sum_{n=m+1}^{\infty} \sum_{\underline{\omega} \in \widetilde{G}_n} \left(-q\tau\bigl(B(\underline{\omega})\bigr) + \alpha \sum_{j=1}^{n-1} \log \lambda\bigl(\sigma^j \omega\bigr) \right),\end{aligned}$$

where ω is any point inside the respective cylinder $[\underline{\omega}]$. Now, using specification and the definition of the delay δ_n, we obtain

$$\begin{aligned}\mathcal{M}_{\xi, G_m}(\alpha, q, F \setminus F_c) \leqslant \bar{c}(\lambda_{\min})^{-\alpha} &\sum_{k=1}^{m+n_0} Z_m(-q + \alpha \log \lambda, \Omega_m) \\ &+ \bar{c} \sum_{n=m+1}^{\infty} \exp(q\delta_n) Z_n(-q + \alpha \log \lambda, \Omega_n),\end{aligned}$$

where $\lambda_{\min} := \min\{\lambda(\omega) \colon \omega \in \Omega\}$, and Z_n is the "partition function" defined in Section 2.4.

It is a general result that

$$Z_n(\psi, \Omega') \leqslant p\, \mathrm{e}^{n P_{\Omega'}(\phi)},$$

for any Hölder continuous potential $\psi : \Omega_p \to \mathbb{R}$ and each specified subshift $\Omega' \subset \Omega_p$. On the other hand, for each $\eta > 0$ and $\alpha = \eta + \sup_n \alpha_c(q, \xi, F_n)$, we have

$$P_{\Omega_n}(-q + \alpha \log \lambda) \leqslant \eta \log(\max_{\Omega_n} \lambda).$$

With this, and hypothesis (H3) we obtain that

$$\mathcal{M}_{\xi,G_m}(\alpha, q, F \setminus F_c) \leqslant \bar{c}(\lambda_{\min})^{-\alpha} \frac{p}{1 - \max_{\Omega_m} \lambda^\eta} + \bar{c} p \sum_{n=1}^{\infty} \exp\bigl(q\delta_n + \eta \log \lambda^{(n)}\bigr) < \infty,$$

for every $m \in \mathbb{N}$. This implies that, for each $\eta > 0$ and $\alpha = \eta + \sup_n \alpha_c(q, \xi, F_n)$, $\mathcal{M}_\xi(\alpha, q, F \setminus F_c) < \infty$, from which we deduce

$$\alpha_c(q, \xi, F \setminus F_c) \leqslant \sup_n \alpha_c(q, \xi, F_n).$$

In this way, the proof of the theorem is finished. □

EXAMPLE 6.2. A minor change in Example 6.1 that makes the critical set to contain an orbit makes impossible to apply Lemma 6.1.

Consider the mapping $\hat{f} : I_0 \cup I_1 \to [0, 1]$ such that

(F1) $\hat{f}$ is strictly increasing and twice continuously differentiable in each branch I_0 and I_1, and it is such that $\hat{f}(I_0) = \hat{f}(I_1) = [0, 1]$.

(F2) $\hat{f}'(x) > 1$ for all $x \in I_1 \cup (I_0 \setminus \{0\})$, $\hat{f}'(0) = 1$, and $f''(x) \geqslant 0$ for all $x \in I_0 \cup I_1$.

See the sketch in Figure 6.2. The repeller associated to this map is the set

$$\widehat{F} = \bigl\{x \in I_0 \cup I_1 \colon \hat{f}^n(x) \in I_0 \cup I_1 \ \forall n \in \mathbb{N}\bigr\}.$$

Let $\hat{g}_0 \equiv (\hat{f}|_{I_0})^{-1}$ and $\hat{g}_1 = (\hat{f}|_{I_1})^{-1}$. These are the inverses of the branches of $\hat{f}$, which are contractions. We repeating all the arguments of the previous ex-

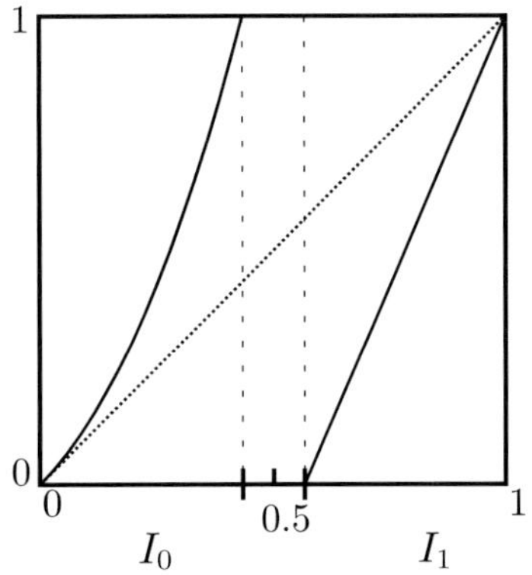

Figure 6.2.

ample, we can defining the corresponding conjugacy $\hat{\chi}$, and potential $\hat{\lambda}$. In this case, the critical set $\hat{\Lambda}_c := \{00000\ldots\}$ is an orbit. To compute the spectrum of dimensions for Poincaré recurrences for $\widehat{F}$, we will use the approximation from below.

For this specific example consider the following approximating sets

$$S_n := \{\omega \in \mathbb{Z}_2^{\mathbb{N}} \colon (\omega_k, \omega_{k+1}, \ldots, \omega_{k+n}) \neq (0, 0, \ldots, 0)\ \forall\, k \in \mathbb{N}\},$$

defining specified subshifts $(S_n\sigma)$ approaching (S, σ). Since the critical set $\widehat{F}_c := \hat{\chi}(\hat{\Lambda}_c)$ contains only a point, its Hausdorff dimension is zero. Hence, hypothesis (H1) is satisfied. On the other hand, it is clear that any sequence $\omega \neq 00000\ldots$ is such that $(\omega_0, \omega_1, \ldots, \omega_{n-1})$ is S_n-admissible for some $n \in \mathbb{N}$, therefore (H2) also holds. In this example, for each $n \in \mathbb{N}$, $G_n = \mathbb{Z}_2^n \setminus \{(0, 0, \ldots, 0)\}$ and $\widetilde{G}_{n+1} = \{(0, 0, \ldots, 0, 1)\}$. It is straightforward to verify that $\delta_n = 0$. Since $\hat{f}$ is a convex function on each branch, then

$$\lambda^{(n)} = \left(\hat{g_0}' \circ \chi\left((0^{n-1}1)^{\infty}\right)\right)^n$$

for every n sufficiently large. Hence, the condition (H3) becomes

$$\sum_{n=1}^{\infty} n \log\left(\hat{g_0}' \circ \hat{\chi}\left((0^{n-1}1)^{\infty}\right)\right) < \infty.$$

Let $a_0 > 1/2 + \delta$ be such that $\hat{f}(a_0) = 0$, and for each $n \in \mathbb{N}$ let $a_n = \hat{g}_0^n(a_0)$. It is easy to see that $\hat{\chi}^{-1}(a_n) = 0^n 10^{\infty}$. In this example, the conjugacy $\hat{\chi}$ maps lexicographic order in $\mathbb{Z}_2^{\mathbb{N}}$ to the real-line order in F. Since $\hat{f}'$ is increasing in each one of the branches I_0 and I_1, then

$$\sum_{n=1}^{\infty} n \log\left(\hat{g_0}'(a_n)\right) < \infty.$$

It is enough to ensure condition (H3). For that we assume that $\hat{f}'(a_n) \geqslant \exp(n^{-\beta})$, for each $n \in \mathbb{N}$ and for some $\beta > 0$.

Chapter 7

The Spectrum for a Sticky Set

Let us adopt the assumptions of Section 3.5, i.e., a sticky set F, the phase space of the dynamical system (F, f), is a result of a strong Moran geometric construction, so that inequalities (3.4) and (3.5) are satisfied and (F, f) is topologically conjugate to a multipermutative system (T, Ω_p). Every minimal multipermutative system is topologically conjugate to the adding machine (Ω_p, T). Theorem 2.1 tells us about universality of sticky sets in the case when the sticky riddle requires the same number of symbols for every level.

7.1. The spectrum of dimensions for Poincaré recurrences

Since $h_{\text{top}}(T) = 0$, then the gauge function $\xi(t)$ should be different from e^{-t}. We know that if a multipermutative system is minimal, the time needed to come back to a cylinder of the length n is exactly p^n. It allows us to guess that the right gauge function is $\xi(t) = 1/t$. So, we find the spectrum $\alpha_c(q, \xi, \Lambda)$ for $\xi(t) = 1/t$.

THEOREM 7.1. *Assume that F is modeled by the full shift (Ω_p, σ) and satisfies the gap condition. Let the system (F, f) be topologically conjugate to a minimal multipermutative system (Ω_p, T). Then, for $\xi(t) = 1/t$ and the parameter region $q \geqslant 0$ and $\alpha \geqslant 0$, the spectrum $\alpha_c(q, \xi)$ is the solution of the equation*

$$P_{\Omega_p}(\alpha \log \lambda) = q \log p. \tag{7.1}$$

The dimension for Poincaré recurrences q_0 is equal to 1.

Thus, we see that again for $q = 0, \alpha_c(0, \xi, \Lambda) = \dim_H F$, the Hausdorff dimension of set F. Moreover, if $\alpha = 0$, the equation becomes: $h_{\text{top}}(v|F) = q \log p$, where $v = \xi \circ \sigma \circ \xi^{-1}$, and, since $h_{\text{top}}(v|F) = \log p$, then $q_0(\xi) = 1$. This result is completely consistent with the observation that $\tau([\omega_0, \dots, \omega_{n-1}]) = p^n$.

PROOF. For an arbitrary minimal multipermutative system the proof is, in fact, the same as for its conjugate system (Ω_p, T). For the sake of definiteness, we prove the theorem for the system (Ω_p, T). Therefore, we will proceed as in the proof of Theorem 5.1 under the assumption of a gap condition. The first step in the proof is the technical Lemma 5.3 that is adapted for multipermutative systems, using the gauge function $\xi(t) = 1/t$, as follows. □

LEMMA 7.1. *Let $F \subset \mathbb{R}^d$ have controlled packing of cylinders with exponent a. Let G be a finite or countable cover of F by open balls. Then there is a positive constant C_0 such that*

$$\mathcal{M}_\xi(\alpha, q, G) \geqslant C_0 \sum_{[\underline{\omega}]\in \mathrm{CMax}(G)} \exp\bigl(-q|\underline{\omega}|\log(p)\bigr)|\underline{\omega}|^{-a} \prod_{j=0}^{|\underline{\omega}|-1} \lambda\bigl(\sigma^j\omega\bigr)^\alpha \tag{7.2}$$

provided that $\alpha \geqslant 0$ and $q \geqslant 0$.

PROOF. The proof is similar to the one for Lemma 5.3. The inequality (5.6) is valid. Moreover, for all $[\underline{\omega}] \in \mathrm{CMax}(B)$ we have $\tau(B) \leqslant \tau([\underline{\omega}])$. By minimality $\tau([\underline{\omega}]) = p^{|\underline{\omega}|}$. Inequality (3.6) implies that

$$\bigl|\log\bigl|\chi([\underline{\omega}])\bigr|\bigr| \leqslant |\log \underline{d}| + |\underline{\omega}|\,|\log \lambda_{\min}|.$$

Therefore,

$$|B|^\alpha \xi\bigl(\tau(B)\bigr)^q \geqslant C_3 \sum_{[\underline{\omega}]\in \mathrm{CMax}(B)} \exp\bigl(-q|\underline{\omega}|\log(p)\bigr)\bigl|\chi([\underline{\omega}])\bigr|^\alpha |\underline{\omega}|^{-a},$$

where $C_3 = C_2/(|\log \lambda_{\min}| + |\log \underline{d}|)^a$. Finally, since $q \geqslant 0$

$$\sum_{B\in G} \sum_{[\underline{\omega}]\in \mathrm{CMax}(B)} \exp\bigl(-q|\underline{\omega}|\log(p)\bigr)\bigl|\chi([\underline{\omega}])\bigr|^\alpha |\underline{\omega}|^{-a}$$

$$\geqslant \underline{d}^\alpha \sum_{[\underline{\omega}]\in \mathrm{CMax}(G)} \exp\bigl(-q|\underline{\omega}|\log(p)\bigr)|\underline{\omega}|^{-a} \prod_{j=0}^{|\underline{\omega}|-1} \lambda\bigl(\sigma^j\omega\bigr)^\alpha$$

and the result follows with $C = C_3\,\underline{d}^\alpha$. □

Now we obtain estimates from above and from below. Let us remind that we are in the case $\xi(t) = 1/t$.

LEMMA 7.2. *For a fixed $q \geqslant 0$ the set function $\mathcal{M}_\xi(\alpha, q) = 0$ for every α such that $P_{\Omega_p}(\alpha\log(\lambda)) < q\log p$.*

PROOF. As in the proof of Lemma 5.1, for a given $\varepsilon > 0$ let $\mathcal{C} \in \mathcal{B}_\varepsilon$ be a cover of F by balls with the property that for any $B \in \mathcal{C}$, $B \cap F = \chi[\underline{\omega}]$ and $|\underline{\omega}| =: n_\varepsilon$

(independent of B). Trivially $n_\varepsilon \to \infty$ as $\varepsilon \to 0$. Again

$$\begin{aligned}\mathcal{M}_\xi(\alpha, q, \varepsilon) &\leqslant \bar{c} \sum_{\substack{[\underline{\omega}]\in S\\ |\underline{\omega}|=n_\varepsilon}} \exp\left(-qn_\varepsilon \log p + \alpha \sum_{j=0}^{n_\varepsilon-1} \log \lambda\big(\sigma^j \omega\big)\right)\\ &\leqslant \bar{c} Z_{n_\varepsilon}\big(-q \log(p) + \alpha \log(\lambda), \Omega_p\big). \end{aligned} \tag{7.3}$$

It follows that $\mathcal{M}_\xi(\alpha, q) = 0$ provided $P_{\Omega_2}(\alpha \log \lambda) < q \log p$. □

LEMMA 7.3. *For a fixed $q \geqslant 0$ the set function $\mathcal{M}_\xi(\alpha, q) = \infty$ for every $\alpha \geqslant 0$ such that $P_{\Omega_p}(\alpha \log(\lambda)) > q \log p$.*

PROOF. Let G be a cover of F by open balls of radius less or equal to ε, with ε small enough to ensure that the inequality (3.27) holds. Thus, by Lemma 7.1 ($a = 1$ in our case, because of Lemma 3.2)

$$\begin{aligned}\mathcal{M}_\xi(\alpha, q, G) \geqslant C_0 &\sum_{[\underline{\omega}]\in \mathrm{CMax}(G)} \exp\Bigg(-q|\underline{\omega}| \log p + \log |\underline{\omega}|\\ &+ \alpha \sum_{j=0}^{|\underline{\omega}|-1} \log \lambda\big(\sigma^j \omega\big)\Bigg). \end{aligned} \tag{7.4}$$

Let $n_\varepsilon = \min\{|\underline{\omega}|\colon [\underline{\omega}] \in \mathrm{CMax}(G)\}$. Since

$$\frac{1}{|\underline{\omega}|} \log |\underline{\omega}| \leqslant \frac{1}{n_\varepsilon} \log n_\varepsilon$$

for all $[\underline{\omega}] \in \mathrm{CMax}(G)$ we get

$$\begin{aligned}\mathcal{M}(\alpha, q, G) \geqslant C &\sum_{[\underline{\omega}]\in \mathrm{CMax}(G)} \exp\Bigg(-|\underline{\omega}|\big(q \log(p) + \log(n_\varepsilon)/n_\varepsilon\big)\\ &+ \alpha \sum_{j=0}^{|\underline{\omega}|-1} \log \lambda\big(\sigma^j \omega\big)\Bigg)\\ &= C\mathcal{Z}\big(q \log(p) + \log n_\varepsilon/n_\varepsilon, \phi, \mathrm{CMax}(G)\big), \end{aligned} \tag{7.5}$$

where $\mathcal{Z}$ is the "statistical sum" in the dimensional definition (Section 2.4.1) of the topological pressure for the potential $\phi(\omega) = \alpha \log \lambda(\omega)$. From this point on, the proof is the same as the one for Lemma 5.2. □

We prove in Chapter 8 a generalization of the theorem to the case of a multi-permutative minimal system.

It is amazing how just the change of the gauge function allowed one to adjust the proof for hyperbolic repeller to this non-chaotic case here. In fact, this change means that we are calculating the return times in a special scale in which they behave in a similar way to the ones for hyperbolic repellers. Such a similarity between a purely chaotic and a completely non-chaotic system could have a deeper nature than we observed in this chapter. For the moment it is just a guess, so we will not speculate about it (some results of such a type for a very specific case can be found in [125]).

Chapter 8

Rhythmical Dynamics

We consider a dynamical system (X, T), where X is compact metric space with a distance d and T is a continuous map. In a dynamical system with discrete time a point x in phase space changes its position at integer moments of time, $i = 1, 2, 3, \ldots$. But it can be easily imagined that the generating map acts at not necessarily integer moments, $t_1, t_2, \ldots t_i \ldots$, which depend on the position of the point x_i and maybe on integer time i. Simple examples of such a situation are Poincaré maps for some flow, induced maps, etc. We provide below another nontrivial example of such a case. In these examples we take into account not only the position of the point in phase space, but also the temporal interval between two successive occurrences of the generating map. In other words, we deal with *rhythmical dynamics*. Such a dynamics is much more rich that the familiar one, since a large amount of the information can be hidden in return times. It was supposed [105] that neural networks take into account not only spatial but temporal information as well. In [86] a symbolic dynamics for a stochastic layer has been suggested, but intervals of time for different states were not justified. In fact, [86] deals with rhythmical dynamics.

It is then natural to study Poincaré recurrences for rhythmical dynamical systems. The present chapter considers rhythmical dynamical systems modeled by some symbolic systems. We study spectra of dimensions for Poincaré recurrences and derive some general formulas for two models. One of them describes a situation for Poincaré maps, another one deals with multipermutative systems and sticky sets. In the first case the temporal interval between two successive iterations depends only on the position of the point x_i; in the second one it depends only on the integer time i. Thus, we study two extreme cases of rhythmical dynamics.

8.1. Set-up

Assume that there exist a continuous nonnegative function $\phi : X \times \mathbb{N} \to \mathbb{R}$ that is responsible for the rhythm, i.e., for any point $x_i = T^i x_0$, the temporal interval between x_i and x_{i+1} is $\phi(x_i, i)$. Generally, the function ϕ could depend on the past of the point x_i as well, but we consider here the simplest situation.

Assume that there exists a finite invariant partition ξ of X and denote by ξ_n the dynamical refinement of ξ, that is: $\xi_n := \bigvee_{j=0}^{n-1} T^{-j}\xi$, where $\xi_0 := \xi$. The atom of the refined partition ξ_n that contains x is denoted by $\xi_n(x)$ and will be referred to as the n-cylinder about x. We assume the following hypothesis are satisfied:

(H_1) atoms of ξ_n are pairwise disjoint;
(H_2) the intersection $\bigcap_{n\geqslant 0} \xi_n(x) = x$ for every $x \in X$;
(H_3) the distance d can be represented as follows: there exists a continuous function $u : X \to (0, \infty)$ such that $d(x, y) = e^{-u(\xi_n(x))}$ whenever $y \in \xi_n(x)$ and $y \notin \xi_{n+1}(x)$, where

$$u\bigl(\xi_n(x)\bigr) := \sup_{k\leqslant n} \sup_{z\in\xi_n(x)} \bigl(u(z) + u(Tz) + \cdots + u\bigl(T^{k-1}z\bigr)\bigr),$$
$$n = 1, 2, \ldots.$$

The case of symbolic systems was described in Section 2.1 already. Quite similarly, here we choose $u(x) = -\log\lambda(x)$ which is a constant on every atom of ξ_0 and is bounded, i.e., there are positive constants $\underline{\lambda} < \bar{\lambda} < 1$ such that $\underline{\lambda} \leqslant \lambda(x) \leqslant \bar{\lambda}$, for every $x \in X$. Then

$$d(x, y) = \prod_{\ell=0}^{n-1} \lambda\bigl(T^\ell x\bigr) \quad \text{and} \quad \operatorname{diam} \xi_n(x) = \prod_{\ell=0}^{n-1} \lambda\bigl(T^\ell x\bigr), \tag{8.1}$$

where $n = \max\{i\colon\ y \in \xi_i(x)\}$.

It is not difficult to see that every open ball is a cylinder $\xi_n(x)$ for some n and x. Remind that a ball of radius ε centered at y is by definition

$$B_\varepsilon(y) := \bigl\{z \in X\colon\ d(z, y) < \varepsilon\bigr\}.$$

For details, see Section 2.1.

8.2. Dimensions for Poincaré recurrences

8.2.1. The case of an autonomous rhythm function ϕ

We consider here a special situation where $\phi(x_i, i) \equiv \phi(x_i)$. We will follow the scheme of the work [3] and take into account that the first return time in the ball $B_\varepsilon(x)$ is not $k := \min\{t \geqslant 1\colon\ T^t(x) \in B_e(x)\}$ but $\sum_{j=0}^{k-1} \phi(T^j x)$.

For an open ball $B \subset X$ let the Poincaré recurrence be defined by

$$\tau(B) = \inf\bigl\{\tau_B(x)\colon\ x \in B\bigr\},$$

where

$$\tau_B(x) = \sum_{j=0}^{k_B(x)-1} \phi\bigl(T^j x\bigr)$$

is the first continuous return time of $x \in B$ and

$$k_B(x) := \min\{t \geqslant 1\colon\ T^t(x) \in B\},$$

the first discrete return time of $x \in B$. The discrete recurrence time of B is defined to be $k(B) := \inf\{k_B(x)\colon\ x \in B\}$. We assume that the rhythm function ϕ is Hölder continuous on X, with exponent $\beta > 0$, i.e., there is $K > 0$ such that

$$|\phi(x) - \phi(y)| \leqslant K\, d(x, y)^\beta \tag{8.2}$$

for every $x, y \in X$. We further assume that ϕ is strictly positive and bounded on X, i.e., there exist positive constants $\underline{\varphi}, \overline{\varphi} > 0$ such that $\underline{\varphi} \leqslant \phi(x) \leqslant \overline{\varphi}$ for every $x \in X$.

For each $A \subset X$, denote by $\mathcal{B}_\varepsilon(A)$ the class of all finite or countable covers of A by balls of diameter less than or equal to ε. Given $G \in \mathcal{B}_\varepsilon(A)$ and $\alpha, q \in \mathbb{R}$, in the general Carathéodory construction, consider the statistical sum $\mathcal{M}_\eta(\alpha, q, \varepsilon, G, A)$ with a real nonnegative function $\eta\colon \mathbb{R} \to \mathbb{R}$ such that $\eta(t) \to 0$ as $t \to \infty$. Below we will consider the functions $\eta(t) = e^{-t}$ and $\eta(t) = 1/t$.

Let us recall that for a fixed q the limit

$$m_\eta(\alpha, q, A) = \lim_{\varepsilon \to 0} \mathcal{M}_{\eta,\varepsilon}(\alpha, q, \varepsilon, A)$$

has an abrupt change from infinity to zero as one varies α from minus infinity to infinity and that the unique critical value

$$\alpha_c(q, \eta) := \alpha_c(q, \eta, A) = \sup\{\alpha\colon\ m_\eta(\alpha, q, A) = \infty\} \tag{8.3}$$

is the spectrum of dimensions for Poincaré recurrences, specified by the function η.

We consider a dynamical system (X, T), where $X \subset \mathbb{R}^d$ is a Cantor set whose geometric construction is described symbolically *à la* Moran and T has the specification property and positive topological entropy (for which we use $\eta(t) = e^{-t}$ in (4.3)). Details are given in Section 8.3.1.

8.2.2. *The case of non-autonomous rhythm function ϕ*

We assume that there exists a function $\phi(x_i, i) \equiv \phi_i$ such that for any n-cylinder B, $\tau(B) = \sum_{i=0}^{n-1} \phi_i$. Such an exotic situation occurs when we deal with multipermutative minimal systems (for which we use $\eta(t) = 1/t$ in (4.3)). The spectrum is defined in the same way as in Section 4.2 of Chapter 4. Details are given in Section 8.3.2.

8.3. The spectrum of dimensions

8.3.1. Autonomous ϕ

It follows from the conjectures (H_1) and (H_2) that the map T is topologically conjugate to a subshift (Ω, σ) (see Section 2.1). In our situation the number of symbols p coincides with the number of atoms in ξ_0. Let us label atoms of ξ_0 by symbols $\{0, 1, \ldots, p-1\}$: $\xi_0 = \{\xi_0^0, \xi_0^1, \ldots, \xi_0^{p-1}\}$. Then by definition $\omega = \omega_0\omega_1 \ldots \omega_{n-1} \ldots \in \Omega$ iff there exists $x \in X$ such that $T^j x \in \xi_0^{\omega_j}$ for every $j \geqslant 0$. The conjugacy is the coding map $\chi : \Omega \to X$, $\chi(\omega) = \bigcap_{j\geqslant 0} T^{-j}\xi_0^{\omega_j}$.

The subshift Ω is assumed to have the specification property. Therefore, there exists an integer n_0 such that for all n-cylinder $[\underline{\omega}]$ the "standard" first-return time $k([\underline{\omega}]) = \inf_{w'\in[\omega]}\{k \geqslant 1\colon \sigma^k w' \in [\underline{\omega}]\}$ satisfies that $k([\underline{\omega}]) \leqslant |\underline{\omega}| + n_0$.

THEOREM 8.1. *Let the system (X, T) be topologically conjugate to a subshift (Ω, σ) with the specification property and positive topological entropy. Then, for $\eta(t) = e^{-t}$ and the parameter region $q \geqslant 0$ and $\alpha \geqslant 0$, the spectrum $\alpha_c(q)$ is the solution of the equation*

$$P_\Omega(-q\phi + \alpha \log \lambda) = 0 \tag{8.4}$$

where $P_\Omega(-q\phi + \alpha \log \lambda)$ is the topological pressure of the function $-q\phi + \alpha \log \lambda$ on the set Ω with respect to σ. The dimension q_0 for Poincaré recurrences coincides with the root of Bowen equation $P_\Omega(q\phi) = 0$.

Let us remark that the standard dynamics is recovered for $\phi \equiv 1$. In this case, because of (2.11), $\alpha_c(q)$ is the root of nonhomogeneous Bowen equation $P_S(\alpha \log \lambda) = q$. This formula was obtained in Theorem 5.1.

PROOF. First we prove that $\mathcal{M}_\eta(\alpha, q, X) = \infty$, whenever $\alpha < \alpha_c(q)$. For an arbitrary cover of X, $\mathcal{C} \in \mathcal{B}_\varepsilon(X)$, we write

$$\mathcal{M}_{\eta,G}(\alpha, q, X) = \sum_{B\in G} e^{-q\tau_B(x_*(B))+\alpha\sum_{j=0}^{n_B-1}\log\lambda(T^j x)}, \tag{8.5}$$

where $x_*(B) \in B$ is such that $\tau(B) = \tau_B(x_*(B))$. For each $B \in \mathcal{C}$, let $x_B \in B$ be periodic with minimal period k_B. It is clear that $\tau_B(x_B) \geqslant \tau_B(x_*(B))$. Then,

$$\mathcal{M}_{\eta,G}(\alpha, q, X) \geqslant \sum_{B\in G} e^{-q\sum_{j=0}^{k_B-1}\phi(T^j x_B)+\alpha\sum_{j=0}^{n_B-1}\log\lambda(T^j x_B)}. \tag{8.6}$$

Due to the specification property of T, we have that $k_B \leqslant n_B + n_0$, for every $B \in \mathcal{C}$. Thus,

$$\mathcal{M}_{\eta,G}(\alpha, q, X) \geqslant e^{-qn_0\bar{\varphi}} \sum_{B\in G} e^{\sum_{j=0}^{n_B-1}(-q\phi(T^j x_B)+\alpha\log\lambda(T^j x_B))}. \tag{8.7}$$

Let $\alpha < \alpha_c(q)$. Then, for any choice of $\mathcal{C} \in \mathcal{B}_\varepsilon(X)$ the right hand side of (8.7) diverges as $\varepsilon \to 0$.

The proof of theorem is completed by proving that $\mathcal{M}_\eta(\alpha, q, X) < \infty$, whenever $\alpha > \alpha_c(q)$. Consider a finite cover $\mathcal{C} = \mathcal{C}_n \in \mathcal{B}_\varepsilon$ such that each $B \in \mathcal{C}_n$ has symbolic length n. As $\varepsilon \to 0$, $n \to \infty$. For this particular cover we have that

$$\mathcal{M}_{\eta,\varepsilon}(\alpha, q, X) \leqslant \sum_{B \in G} e^{-q\tau(B)+\alpha\sum_{j=0}^{n-1}\log\lambda(T^j x)}. \tag{8.8}$$

Denote by $P_k \subset \mathcal{C}$ the subcollection of balls in $\mathcal{C}$ having first discrete return time k. Then, by the specification property of T we have that

$$\begin{aligned}
&\mathcal{M}_{\eta,\varepsilon}(\alpha, q, X) \\
&\quad \leqslant \sum_{k=1}^{n+n_0} \sum_{B \in P_k} e^{-q\tau(B)+\alpha\sum_{j\geqslant 0}^{k-n_0-1}\log\lambda(T^j x)} \\
&\quad \leqslant \underline{\lambda}^{-n_0\alpha} \sum_{k=1}^{n+n_0} \sum_{B \in P_k} e^{-q\tau(B)+\alpha\sum_{j=0}^{k-1}\log\lambda(T^j x)}.
\end{aligned}$$

For cylinder $B \in P_k$, let the integer k_{*B} and the point $x_{*B} \in B$ be such that $\tau(B) = \sum_{j=0}^{k_{*B}-1}\phi(T^j x_{*B})$. Then, since $k_{*B} \geqslant k$,

$$\begin{aligned}
&\mathcal{M}_{\eta,\varepsilon}(\alpha, q, X) \\
&\quad \leqslant \underline{\lambda}^{-n_0\alpha} \sum_{k=1}^{n+n_0} \sum_{B \in P_k} e^{-q\sum_{j=0}^{k-1}\phi(T^j x_{*B})+\alpha\sum_{j=0}^{k-1}\log\lambda(T^j x)} \\
&\quad \leqslant \underline{\lambda}^{-n_0\alpha} \sum_{k=1}^{\infty} \sum_{B \in \mathcal{C}_k} e^{\sum_{j=0}^{k-1}(-q\phi(T^j x_* B)+\alpha\log\lambda(T^j x_{*B}))},
\end{aligned} \tag{8.9}$$

where $\mathcal{C}_k$ is a cover of X by all nonempty cylinders of length k. For $\alpha > \alpha_c(q)$ the upper bound in (8.9) remains finite in the limit $\varepsilon \to 0$. Indeed, in this limit the first sum in (8.9) behaves as $\exp(kP_S(-q\phi + \alpha\log\lambda))$ with topological pressure $P_S(-q\phi + \alpha\log\lambda) < 0$ such that the second sum in k converges. $\square$

8.3.2. Non-autonomous ϕ

Let $p_* := (p_0, p_1, \ldots, p_i, \ldots)$ be a sequence of integers, $p_i \geqslant 2, i = 0, 1, 2, \ldots$. Let

$$F : \prod_{i\geqslant 0}\mathbb{Z}_{p_i} \to \prod_{i\geqslant 0}\mathbb{Z}_{p_i}$$

be a multipermutative map for which the sequence of integers p_i is "regular" enough so that the following limit exists,

$$\langle \log p \rangle := \lim_{n\to\infty} \frac{1}{n} \sum_{i=0}^{n} \log p_i . \tag{8.10}$$

An example of a multipermutative system is the p-adic adding machine, described in Section 2.3. In that case the symbolic system is minimal.

THEOREM 8.2. *Let the system (X, T) be topologically conjugate to a minimal multipermutative system (Ω_{p_*}, F). Then, for $\xi(t) = 1/t$ and the parameter region $q \geqslant 0$ and $\alpha \geqslant 0$, the spectrum $\alpha_c(q, \xi)$ is the solution of the equation*

$$P_{\Omega_p}(\alpha \log \lambda) = q \langle \log p \rangle, \tag{8.11}$$

where $P_{\Omega_{p*}}(\alpha \log \lambda)$ is the topological pressure of the function $\alpha \log \lambda$ on the set Ω_{p*} with respect to σ. In this case the dimension for Poincaré recurrences, q_0, specified by $\xi(t) = 1/t$, is equal to $h_{\text{top}}/\langle \log p \rangle$, where h_{top} is the topological entropy of σ on the (generally non-invariant) set Ω_{p*} (see Section 4.1 for the definition).

The formula (8.11) was proved in Theorem 7.1 for the case $p_i = \text{constant} = p$.

COROLLARY 8.1. $q_0 = 1$.

PROOF OF THEOREM 8.2. We consider covers of Ω_{p_*}, $\mathcal{C}_\varepsilon$, by cylinders of diameter $\varepsilon > 0$ sufficiently small such that for every $\delta > 0$

$$\left| \frac{1}{n} \sum_{i=1}^{n-1} \log p_i - \langle \log p \rangle \right| < \delta, \tag{8.12}$$

for any $n \geqslant n_\varepsilon$. Notice that $\varepsilon \to 0$ as $\delta \to 0$.

First we prove that $\mathcal{M}_\eta(\alpha, q, \Omega_{p*}) < \infty$ whenever $q \langle \log p \rangle > P_{\Omega_{p*}}(\alpha \log \lambda)$. For any $\delta > 0$ and the class of covers considered we have that

$$\begin{aligned} \mathcal{M}_{\eta,\varepsilon}(\alpha, q, \Omega_{p*}) &\leqslant \sum_{B\in\mathcal{C}_\varepsilon} e^{-q\sum_{j=0}^{|B|-1} \log p_j + \alpha \sum_{j=0}^{|B|-1} \log(\lambda\sigma^j\omega)} \\ &\leqslant \sum_{B\in\mathcal{C}_\varepsilon} e^{-q|B|(\langle \log p\rangle - \delta) + \alpha\sum_{j=0}^{|B|-1} \log(\lambda\sigma^j\omega)} \end{aligned} \tag{8.13}$$

where we are using the gauge function $\eta(t) = 1/t$. The inequality (8.13) follows from (8.12) since $|B| \geqslant n_\varepsilon$ for every $B \in \mathcal{C}_\varepsilon$. The sum in (8.13) has the form of the statistical sum in the dimension-like definition of topological pressure (2.17).

Then, $\mathcal{M}_{\eta,\varepsilon}(\alpha, q, \Omega_{p*}) < \infty$ whenever

$$q\big(\langle\log p\rangle - \delta\big) \geqslant P_{\Omega_{p*}}(\alpha \log\lambda).$$

Since δ is arbitrary we let it vanish to get the announced result.

Finally we prove that $\mathcal{M}_\eta(\alpha, q, \Omega_{p*}) = \infty$ whenever $q\langle\log p\rangle \leqslant P_{\Omega_{p*}}(\alpha\log\lambda)$. For any $\delta > 0$, let $\mathcal{C}_\varepsilon$ be a cover of Ω_{p_*} such that (8.12) is satisfied with $n = |B|$ for every $B \in \mathcal{C}_\varepsilon$. For any such cover we have that

$$\mathcal{M}_{\eta,\varepsilon}(\alpha, q, \Omega_{p_*}) \geqslant \sum_{B\in\mathcal{C}_\varepsilon} e^{-q|B|(\langle\log p\rangle+\delta)+\alpha\sum_{j=0}^{|B|-1}\log(\lambda\sigma^j\omega)}. \tag{8.14}$$

As $\delta \to 0$ the sum in (8.14) diverges for every $\mathcal{C}_\varepsilon$ (recall that $\varepsilon \to 0$ as $\delta \to 0$) whenever $q(\langle\log p\rangle + \delta) \leqslant P_{\Omega_{p*}}(\alpha\log\lambda)$, i.e., $\mathcal{M}_\eta(\alpha, q, \Omega_{p_*}) = \infty$ whenever $q\langle\log p\rangle \leqslant P_{\Omega_{p*}}(\alpha\log\lambda)$. □

PROOF OF COROLLARY 8.1. The topological entropy h_{top} is the threshold value β_0 in the statistic sum $\sum_{B\in\mathcal{C}_\varepsilon} e^{-\beta|B|}$, where $\mathcal{C}_\varepsilon$ is a cover of Ω_{p_*} by cylinders of length greater than n_ε, $n_\varepsilon \to \infty$ as $\varepsilon \to 0$ (see Section 4.1). Direct calculations show that if $|B| = \text{constant} = n$, then $\sum_{B\in\mathcal{C}_\varepsilon} e^{-\beta|B|} \to 0$, provided that $\beta > \langle\log p\rangle$. It means that $\beta_0 \leqslant \langle\log p\rangle$. Assume that $\beta < \langle\log p\rangle$. Then

$$\sum_{B\in\mathcal{C}_\varepsilon} e^{-\beta|B|} \geqslant \sum_{B\in\mathcal{C}_\varepsilon} e^{-\beta n_e} \geqslant \prod_{i=0}^{n_e-1} p_i e^{-\beta n_e} \to \infty$$

as $\varepsilon \to 0$. Thus, $\beta_0 \geqslant \langle\log p\rangle$. □

PART III

ONE-DIMENSIONAL SYSTEMS

Chapter 9

Markov Maps of the Interval

The formula in Theorem 5.1 for zero-dimensional systems holds for Markov maps of the interval, $T : I \to I$ (see [52]). They were defined in Section 3.1.2. An important feature of Markov maps is that they have a finite Markov partition $\{I_k\}_{k=0}^{p-1}$ by means of which a map $\chi : \Omega_A \to I$ is constructed such that $T \circ \chi = \chi \circ \sigma$. The subshift Ω_A is determined by a transition matrix $A : \{0, \dots, p-1\} \times \{0, \dots, p-1\} \to \{0, 1\}$ with entries $A_{i,j} = 1$ whenever $T(I_i) \supset I_j$ and $A_{i,j} = 0$ otherwise. To better follow this chapter, the reader will find convenient to review Section 2.2 and Section 3.1.2. The graph of a Markov map, with $p = 3$, is shown in Figure 9.1 to highlight main facts. Remark that there are no gaps between the basic intervals I_0–I_1 and I_1–I_2. We will see that the disconnectedness condition in Theorem 5.1 can be relaxed in the case of uniformly hyperbolic systems.

The main result in this chapter is Theorem 9.1, the proof of which has its own interest. Indeed, it shows that the construction of mixing subshifts $\{(\Omega_n, \sigma)\}_{n \geqslant N}$, for some N, of a given Markov chain (Ω_A, σ) discussed in Section 2.2 provides us with a method for approximating a Markov map (T, I) by a sequence of maps restricted on invariant Cantor sets Λ_n, $T|\Lambda_n$, that are conjugated to mixing subshifts of finite type, (Ω_n, σ). The map $\chi|\Omega_n$ provides the conjugacy.

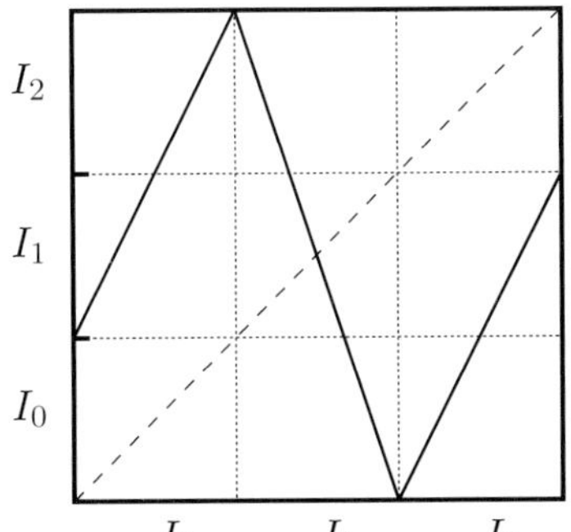

Figure 9.1.

The interest of the method is that it could be used for other situations. It can, for instance, be used to compute the pressure of some non-uniformly hyperbolic Markov maps of the interval, although in this case the conditions of Theorem 9.1 are not sufficient. However, if further conditions on the diameter of cylinders are provided the method of approximations applies.

The main steps in the proof of Theorem 9.1 are to apply Theorem 5.1 to the zero-dimensional approximations and then prove the convergence of the pressure of the sequence of approximations to the pressure of the full Markov map (Proposition 2.5). We should point out that convergence to the full pressure does not follow from the well-known continuity of the pressure with respect to the potential function [108].

9.1. The spectrum of dimensions

THEOREM 9.1. *Let T be a Markov map on an interval I such that the associated topological Markov chain (Ω_A, σ) is mixing. Then, in the domain α, $q \in \mathbb{R}^+$, the spectrum of dimension for Poincaré recurrences $\alpha_c(q, I)$ is the solution of the equation $P_{\Omega_A}(\alpha \log \lambda) = q$.*

Hence, the Poincaré spectrum for mixing Markov maps is again the solution of a nonhomogeneous Bowen equation. In the case of piecewise affine maps whose topological Markov chain is the full shift on p symbols, it is a simple exercise to show that the pressure is given by

$$P_{\Omega_A}(\alpha \log \lambda) = \log \sum_{i=0}^{p-1} \lambda_i^{\alpha},$$

where the number λ_i^{-1} is the inverse of the modulus of $T'|\mathrm{Int}(I_i)$ and I_i are intervals of continuity. For these maps, the spectrum then satisfies the nonhomogeneous Bowen equation

$$\log \sum_{i=0}^{p-1} \lambda_i^{\alpha_c(q)} = q.$$

PROOF OF THEOREM 9.1. We compute the spectrum of (I, T) by taking coverings by images of cylinders of (Ω_A, σ) (which we also call cylinders) rather than balls. We denote the outer measure $\mathcal{M}(\alpha, q, Y, \mathcal{B})$ and the spectrum $\alpha_c(q, Y, \mathcal{B})$ when using covers by balls, and denote them by $\mathcal{M}(\alpha, q, Y, \mathcal{C})$ and by $\alpha_c(q, Y, \mathcal{C})$ when using covers by cylinders. In the case of a map on a Cantor set $\Lambda \subset \mathbb{R}$ conjugated to a specified subshift, it was proved in [2] that

$$\alpha_c(q, \Lambda) := \alpha_c(q, \Lambda, \mathcal{B}) = \alpha_c(q, \Lambda, \mathcal{C}).$$

The corresponding statement in our framework is the following.

PROPOSITION 9.1. *If the topological Markov chain associated to a Markov map is mixing, then in the domain* $\alpha, q \in \mathbb{R}^+$, *we have*

$$\alpha_c\left(q, \bigcup_{n=N}^{\infty} \Lambda_n, \mathcal{C}\right) = \sup_{n \geqslant N} \alpha_c(q, \Lambda_n, \mathcal{C}) = \alpha_c(q, I, \mathcal{B}) =: \alpha_c(q, I)$$

where the invariant subsets Λ_n *are the Cantor sets constructed in Section* 2.2.

The monotonicity of the pressure implies that, for each $n \geqslant N$, we have

$$P_{\Omega_n}\big(\alpha_c(q, I) \log \lambda\big) \geqslant q,$$

and then by Proposition 2.5 we have $P_{\Omega_A}(\alpha_c(q, I) \log \lambda) \geqslant q$. On the other hand, if $\alpha < \alpha_c(q, I)$, then there exists n_α such that $P_{\Omega_n}(\alpha \log \lambda) \leqslant q$, $n \geqslant n_\alpha$, and thus $P_{\Omega_A}(\alpha \log \lambda) \leqslant q$. Finally, the continuity of $P_{\Omega_A}(\alpha \log \lambda)$ with α implies that $P_{\Omega_A}(\alpha_c(q, I) \log \lambda) \leqslant q$, and the theorem follows. □

REMARK 9.1. If we consider a measure with support the full interval, we observe that the set $\bigcup_{n=N}^{\infty} \Lambda_n$ does not contain any typical point. However, Proposition 9.1 states that this set has full dimension $\alpha_c(q, I)$ for every $q \geqslant 0$. In particular, this set has full Hausdorff dimension, $\alpha_c(q = 0, I)$, and full topological entropy, $\alpha_c(h_{\text{top}}, I) = 0$.

PROOF OF PROPOSITION 9.1. We divide the proof in two parts. First we prove

$$\alpha_c\left(q, \bigcup_{n=N}^{\infty} \Lambda_n, \mathcal{C}\right) = \sup_{n \geqslant N} \alpha_c(q, \Lambda_n, \mathcal{C}). \tag{9.1}$$

For any $n \geqslant N$, we have $\Lambda_n \subset \bigcup_{n=N}^{\infty} \Lambda_n$. Since $\mathcal{M}(\alpha, q, \cdot, \mathcal{C})$ is an outer measure, the condition

$$\alpha > \alpha_c\left(q, \bigcup_{n=N}^{\infty} \Lambda_n, \mathcal{C}\right)$$

implies that $\alpha \geqslant \alpha_c(q, \Lambda_n, \mathcal{C})$ and then

$$\alpha_c(q, \Lambda_n, \mathcal{C}) \leqslant \alpha_c\left(q, \bigcup_{n=N}^{\infty} \Lambda_n, \mathcal{C}\right).$$

Consequently

$$\sup_{n \geqslant N} \alpha_c(q, \Lambda_n, \mathcal{C}) \leqslant \alpha_c\left(q, \bigcup_{n=N}^{\infty} \Lambda_n, \mathcal{C}\right).$$

To prove the converse inequality, let $\varepsilon > 0$ and let $k_\varepsilon \in \mathbb{Z}^+$ be such that for any $k > k_\varepsilon$ and for any cylinder $[\omega_0 \dots \omega_k]$, we have $\operatorname{diam}(\chi[\omega_0 \dots \omega_k]) \leqslant \varepsilon$. Consider the finite covering of Λ_n by cylinders of length $k_\varepsilon + 1$. From this covering, we construct another covering, denoted by $C_{n,\varepsilon}$, using the following procedure.

If the Poincaré recurrence $\tau([\omega_0 \dots \omega_{k_\varepsilon}])$ of a given cylinder $[\omega_0 \dots \omega_{k_\varepsilon}]$, is not greater than k_ε, then its image $\chi([\omega_0 \dots \omega_{k_\varepsilon}])$ belongs to $C_{n,\varepsilon}$. Otherwise, consider all the cylinders $[\omega_0 \dots \omega_{k_\varepsilon+1}]$ contained in $[\omega_0 \dots \omega_{k_\varepsilon}]$. For each cylinder, if $\tau([\omega_0 \dots \omega_{k_\varepsilon+1}]) \leqslant k_\varepsilon + 1$, then $\chi([\omega_0 \dots \omega_{k_\varepsilon+1}]) \in C_{n,\varepsilon}$. Otherwise repeat the process.

Because of mixing, there exists an integer m such that for every cylinder $[\omega_0 \dots \omega_k] \subset \Omega_n$ we have $\tau([\omega_0 \dots \omega_k]) \leqslant k + m$. Therefore, the process ends after a finite number of iterations, i.e., $C_{n,\varepsilon}$ is a finite covering. The set $C_\varepsilon = \bigcup_{n=N}^{\infty} C_{n,\varepsilon}$ is a countable covering of $\bigcup_{n=N}^{\infty} \Lambda_n$ by cylinders. We compute the value of $M(\alpha, q, \bigcup \Lambda_n, \mathcal{C})$ for this covering.

Since all the terms in the sum of $M(\alpha, q, \bigcup \Lambda_n, \mathcal{C})$ are non-negative, the series can be computed first by summing over the Poincaré recurrences and then by summing over the cylinders with given Poincaré recurrence. Any cylinder $\chi[\omega_0 \dots \omega_\ell] \in C_\varepsilon$ belongs to $\chi[\omega_0 \dots \omega_{\tau([\omega_0 \dots \omega_\ell])}]$. Therefore we have that

$$
\begin{aligned}
M\left(\alpha, q, \bigcup_{n=1}^{\infty} \Lambda_n, C_\varepsilon\right)
&= \sum_{t=1}^{\infty} \sum_{\chi([\omega_0 \dots \omega_\ell]) \in C_\varepsilon:\ \tau([\omega_0 \dots \omega_\ell]) = t} e^{-qt} \left(\operatorname{diam} \chi\left([\omega_0 \dots \omega_\ell]\right)\right)^\alpha \\
&\leqslant \sum_{t=1}^{\infty} \sum_{\chi([\omega_0 \dots \omega_\ell]) \in C_\varepsilon:\ \tau([\omega_0 \dots \omega_\ell]) = t} e^{-qt} \left(\operatorname{diam} \chi\left([\omega_0 \dots \omega_t]\right)\right)^\alpha .
\end{aligned}
$$

Let C_t be the set of cylinders $\chi[\omega_0 \dots \omega_t] \in \Lambda_t$ such that $\omega_0 = \omega_t$. If a cylinder of C_ε with Poincaré recurrence $t \geqslant N$ belongs to Λ_t, then it belongs to C_t. Moreover, Proposition 2.1 tells us that the number of periodic orbits of period n not belonging to Λ_n is bounded by a constant m, for every $n \geqslant N$. Therefore, the number of cylinders of C_ε with return time $t \geqslant N$ not belonging to Λ_t is bounded by m. Accordingly, we decompose the last sum into three sums

$$
\sum_{t=1}^{\infty} \sum_{\chi([\omega_0 \dots \omega_\ell]) \in C_\varepsilon:\ \tau([\omega_0 \dots \omega_\ell]) = t} e^{-qt} \left(\operatorname{diam} \chi\left([\omega_0 \dots \omega_t]\right)\right)^\alpha = \Sigma_1 + \Sigma_2 + \Sigma_3 .
$$

The first sum

$$
\Sigma_1 = \sum_{t=1}^{N-1} \sum_{\chi([\omega_0 \dots \omega_\ell]) \in C_\varepsilon:\ \tau([\omega_0 \dots \omega_\ell]) = t} e^{-qt} \left(\operatorname{diam} \chi\left([\omega_0 \dots \omega_t]\right)\right)^\alpha ,
$$

can be bounded from above by a number depending only on N because

$$\begin{aligned}&\#\{\chi([\omega_0\dots\omega_\ell])\in C_\varepsilon\colon\ \tau([\omega_0\dots\omega_\ell])<N\}\\&\quad\leqslant\#\{[\omega_0\dots\omega_t]\in\Omega_A\colon\ t\leqslant N\}.\end{aligned}$$

Using the upper bound in condition (3.11), the second sum can be bounded as follows

$$\begin{aligned}\Sigma_2&=\sum_{t=N}^{\infty}\ \sum_{\chi([\omega_0\dots\omega_\ell])\in C_\varepsilon\setminus C_t\colon\ \tau([\omega_0\dots\omega_\ell])=t}e^{-qt}\big(\operatorname{diam}\chi([\omega_0\dots\omega_t])\big)^\alpha\\&\leqslant\sum_{t=N}^{\infty}me^{-qt}(\operatorname{diam}I)^\alpha\lambda_{\max}^{t\alpha},\end{aligned}$$

where $\lambda_{\max}=\max_{\omega\in\Omega_A}\lambda(\omega)<1$. Consequently, Σ_2 is bounded by a number independent of ε if $\max\{\alpha,q\}>0$. The third sum is

$$\begin{aligned}\Sigma_3&=\sum_{t=N}^{\infty}\ \sum_{\chi([\omega_0\dots\omega_\ell])\in C_\varepsilon\colon\ \chi([\omega_0\dots\omega_t])\in C_t}e^{-qt}\big(\operatorname{diam}\chi([\omega_0\dots\omega_t])\big)^\alpha\\&\leqslant e^q\sum_{t=N}^{\infty}\ \sum_{\chi([\omega_0\dots\omega_t])\in C_t}e^{-q(t+1)}\big(\operatorname{diam}\chi([\omega_0\dots\omega_t])\big)^\alpha.\end{aligned}$$

Using again the upper bound of (3.11), the conditions $q\geqslant 0$ and $0<\lambda<1$, one can write

$$\begin{aligned}&\sum_{\chi([\omega_0\dots\omega_t])\in C_t}e^{-q(t+1)}\big(\operatorname{diam}\chi([\omega_0\dots\omega_t])\big)^\alpha\\&\quad\leqslant\rho^\alpha\sum_{i=0}^{p-1}\ \sum_{\omega\in[i]\colon\ \sigma^t\omega=\omega}\ \prod_{j=0}^{t}e^{-q}\big(\lambda(\sigma^j\omega)\big)^\alpha\\&\quad\leqslant\rho^\alpha\sum_{i=0}^{p-1}\big(\mathcal{L}^t_{-q+\alpha\log\lambda,t}1_{[i]}\big)(\omega^i),\end{aligned}$$

for any p-tuple $\{\omega^i\}_{i=0}^{p-1}$ with $\omega^i\in[i]$. Here

$$(\mathcal{L}_{\varphi,t}f)(\omega)=\sum_{\omega'\in\sigma^{-1}(\omega)}\exp\big(\varphi(\omega')\big)f(\omega')$$

is the Ruelle–Perron–Frobenius operator on the continuous functions $f:\Omega_t\to\mathbb{R}$ and $1_{[i]}$ is the characteristic function of the set $[i]$.

Because (Ω_t,σ) is mixing and the potential $-q+\alpha\log\lambda$ is Hölder continuous, the Perron–Frobenius Theorem ensures the existence of an eigenfunction $h_t>0$

such that

$$\mathcal{L}_{-q+\alpha \log \lambda, t}\, h_t = \exp\bigl(P_{\Omega_t}(-q+\alpha \log \lambda)\bigr) h_t$$

where $P_{\Omega_t}(-q+\alpha \log \lambda)$ is the topological pressure associated to the potential $-q+\alpha \log \lambda$ [108].

For each $i \in \{0, \ldots, p-1\}$ we normalize h_t such that $\min\{h_t(\omega)\colon \omega \in [i]\} = 1$ and we denote the normalized eigenfunction by h_t^i. We have $h_t^i \geqslant 1_{[i]}$ and, since the Ruelle operator is positive, it results that $\mathcal{L}_{\varphi,t}^k 1_{[i]} \leqslant \exp(k P_{\Omega_t}(\varphi)) h_t^i$ for all $k \geqslant 1$. Hence, choosing for each i, $\omega^i \in [i]$ such that $h_t^i(\omega^i) = 1$, we obtain

$$\sum_{i=0}^{p-1} \bigl(\mathcal{L}_{\varphi,t}^t 1_{[i]}\bigr)(\omega^i) \leqslant p \exp\bigl(t P_{\Omega_t}(\varphi)\bigr), \tag{9.2}$$

and then

$$\Sigma_3 \leqslant p\rho^\alpha e^q \sum_{t=N}^{\infty} e^{t P_{\Omega_t}(-q+\alpha \log \lambda)}.$$

Now assume that $\alpha = \sup_{t \geqslant N} \alpha_c(q, \Lambda_t, \mathcal{C}) + \delta$ for some $\delta > 0$. Then, for each $t \geqslant N$, we have $-q+\alpha \log \lambda \leqslant -q + \alpha_c(q, \Lambda_t, \mathcal{C}) \log \lambda + \delta \log \lambda_{\max}$. It follows from the definition of the pressure that

$$P_{\Omega_t}(-q+\alpha \log \lambda) \leqslant P_{\Omega_t}\bigl(-q+\alpha_c(q, \Lambda_t, \mathcal{C}) \log \lambda\bigr) + \delta \log \lambda_{\max}.$$

Since Λ_t and Ω_t are topologically conjugated and relation (3.11) holds, then each set Λ_t satisfies the assumptions of Theorem 5.1. It results that $P_{\Omega_t}(-q + \alpha \log \lambda) \leqslant \delta \log \lambda_{\max}$, and consequently

$$\Sigma_3 \leqslant p\rho^\alpha e^q \sum_{t=N}^{\infty} \bigl(\lambda_{\max}^\delta\bigr)^t,$$

i.e., Σ_3 is bounded by a number independent of ε. Since $\alpha > 0$, the sum Σ_2 is also bounded by a number independent of ε. Since ε is arbitrary, we conclude that the condition $\alpha > \sup_{n \geqslant N} \alpha_c(q, \Lambda_n, \mathcal{C})$ implies that

$$\alpha \geqslant \alpha_c\left(q, \bigcup_{n=N}^{\infty} \Lambda_n, \mathcal{C}\right)$$

which proves the equality (9.1).

Let us now prove the equality

$$\sup_{n \geqslant N} \alpha_c(q, \Lambda_n, \mathcal{C}) = \alpha_c(q, I, \mathcal{B}). \tag{9.3}$$

Since $\Lambda_n \subset I$, we have $\alpha_c(q, \Lambda_n, \mathcal{B}) \leqslant \alpha_c(q, I, \mathcal{B})$. From Theorem 5.1 and Proposition 9.1, it follows that

$$\alpha_c\left(q, \bigcup_{n=N}^{\infty} \Lambda_n, \mathcal{C}\right) = \sup_{n \geqslant N} \alpha_c(q, \Lambda_n, \mathcal{C}) \leqslant \alpha_c(q, I, \mathcal{B}),$$

and one inequality is proved.

To prove the converse inequality we consider the set

$$\Lambda_e = I \setminus \bigcup_{n=N}^{\infty} \bigcup_{[\omega_0 \ldots \omega_n] \subset \Omega_A} \partial I_{\omega_0 \ldots \omega_n},$$

where intervals $I_{\omega_0 \ldots \omega_n}$ are the nth dynamical refinement of the basic intervals I_{ω_0} (see Section 3.1.2). Since I and Λ_e only differ by a countable set of points, we have $\alpha_c(q, I) = \alpha_c(q, \Lambda_e)$.

Let $\varepsilon > 0$ and let $k_\varepsilon \geqslant N$ be such that the diameter of the χ-image of any cylinder of length $k_\varepsilon + 1$ is at most ε. Consider the collection

$$B_\varepsilon = \left\{\operatorname{Int}(I_{\omega_0 \ldots \omega_{k_\varepsilon}})\colon [\omega_0 \ldots \omega_{k_\varepsilon}] \subset \Omega_A\right\}.$$

The set B_ε is a cover of Λ_e. Computing M for this cover, we obtain using the relation (3.11)

$$\begin{aligned} M(\alpha, q, \Lambda_e, B_\varepsilon) &\leqslant \rho^\alpha \sum_{[\omega_0 \ldots \omega_{k_\varepsilon}] \subset \Omega_A} e^{-q\tau([\omega_0 \ldots \omega_{k_\varepsilon}])} \prod_{i=0}^{k} \left(\lambda\left(\sigma^i \omega\right)\right)^\alpha \\ &\leqslant \rho^\alpha \sum_{t=1}^{k_\varepsilon + k_0} \sum_{[\omega_0 \ldots \omega_{k_\varepsilon}] \subset \Omega_A} \exp\left(\sum_{i=0}^{t} \varphi\left(\sigma^i \omega\right)\right) \\ &\leqslant \rho^\alpha \sum_{t=1}^{\infty} C e^{t P_{\Omega_A}(\varphi)}, \end{aligned} \tag{9.4}$$

where $\varphi = -q + \alpha \log \lambda$ and k_0 is the mixing time of (Ω_A, σ), i.e., k_0 is the smallest integer number such that the matrix A^{k_0} is positive.

Moreover, as follows from the proof of Proposition 9.1, the condition

$$\alpha \geqslant \alpha_c\left(q, \bigcup_{n=N}^{\infty} \Lambda_n, \mathcal{C}\right) + \delta$$

for some $\delta > 0$ implies that $P_{\Omega_n}(\varphi) \leqslant \delta \log \lambda_{\max}$ for all $n \geqslant N$. Using Proposition 2.5, the previous condition then implies $P_{\Omega_A}(\varphi) \leqslant \delta \log \lambda_{\max}$ and then

$$M(\alpha, q, \Lambda_e, B_\varepsilon) < \overline{M},$$

and $\overline{M}$ does not depend on ε. Since the result holds for any $\varepsilon > 0$, equality (9.3) follows. $\square$

REMARK 9.2. In the proof we used essentially the theory of ordered topological Markov chains, described in Section 2.2, and the assumption that the map is Markov. Nevertheless, we believe that for any piece-wise expanding map of the interval the Bowen equation is still valid. At least we know that the technique used above can be extended to the maps semi-conjugated with sophic symbolic systems satisfying the specification condition.

Chapter 10

Suspended Flows

10.1. Suspended flows over specified subshifts

This chapter is devoted to a special flow over a specified mixing subshift. It inherits, of course, all chaotic properties of the subshift but the behavior of Poincaré recurrences can be very different (see, for instance, [6]). To expose the fact that it still satisfies the Bowen-type equation we exploit essentially the Bowen–Walters' distance and, as the reader will see, it is a nontrivial task.

Let (X,σ) be a subshift of (Ω_N,σ). For the sake of simplicity of notations, in this section we denote by X a subshift and by $x, y, z, \ldots$ points in X. Denote ζ the partition of X into 1-cylinders, and $\zeta^n = \bigvee_{j=0}^{n-1} \sigma^{-j}\zeta$ the dynamical refinements of this partition. Finally, let us denote ζ^0 the trivial partition $\{X, \emptyset\}$.

We assume that (X,σ) is weakly specified, i.e., there exists an integer number $n_0 > 0$ such that for any two cylinders $c \in \zeta^n$ and $c' \in \zeta^m$, and for each integer number $\ell \geqslant n+m+2n_0$, there exists a periodic point of period ℓ such that $x \in c$ and $\sigma^{n+n_0}x \in c'$.

For each $x \in X$ let $\zeta^n(x)$ be the atom of ζ^n containing x. We now remind the definition of the metric on X generating the product topology, introduced in Section 2.1.1. Let $u : X \to (0,\infty)$ be a continuous function, and define the distance $d_X(x,y) = \exp(-u(\zeta^n(x)))$, where $n = \max\{k \in \mathbb{N}\colon\ \zeta^k(x) = \zeta^k(y)\}$ and

$$u\bigl(\zeta^n(x)\bigr) = \max_{z \in \zeta^n(x)} \sum_{j=0}^{n-1} u\bigl(\sigma^j z\bigr).$$

In the sequel we will assume that u is Hölder continuous.

We have already proved in Chapter 2 that d_X is an ultrametric. We have seen that open balls coincides with cylinder sets, i.e., for each $x \in X$ and $\varepsilon > 0$ there exists a unique $n_{x,\varepsilon} \in \mathbb{Z}^+$ such that $B(x,\varepsilon) = \zeta^{n_{x,\varepsilon}}(x)$. And vice versa, for each $x \in X$ and $n \in \mathbb{N}$ there exists $\varepsilon_{x,n} > 0$ such that $\zeta^n(x) = B(x,\varepsilon_{x,n})$, thought this $\varepsilon_{x,n}$ is not unique.

10.1.1. Poincaré recurrences

Let us remind that

$$\tau_\sigma(U) = \min\{k \in \mathbb{N}\colon \sigma^k(U) \cap U \neq \emptyset\}$$

is the Poincaré recurrence of the set $U \subset X$. Note that $\tau_\sigma(U) = \tau_\sigma(\sigma^{-1}U)$ for each $U \subset X$, and that $\tau_\sigma(A) \leqslant \tau_\sigma(B)$ whenever $A \subset B$.

10.1.2. Suspended flow

Let $\phi: X \to (0, \infty)$ be a Hölder continuous function. Consider the interval $[0, \infty)$ with the usual topology, and endow $X \times [0, \infty)$ with the product topology. Define now the equivalence relation $\sim_\phi$ in $X \times [0, \infty)$ as follows. For $(x, t) \in X \times [0, \infty)$, let

$$n(x,t) := \max\left\{n \in \mathbb{N}\colon \sum_{k=0}^{n} \phi(\sigma^k x) \leqslant t\right\},$$

and $s(x,t) := t - \sum_{k=0}^{n(x,t)} \phi(\sigma^k x)$. Then let

$$(x,t) \sim_\phi \left(\sigma^{n(x,t)}x, t - s(x,t)\right),$$

and extend this relation by symmetry and transitivity. The suspended space is the quotient $X^\phi := X \times [0, \infty)/\sim_\phi$. To simplify notation, let us identify each class in X^ϕ with its representative in $\{(x,t)\colon x \in X,\ 0 \leqslant t < \phi(x)\}$.

In the suspended space X^ϕ we define the suspended flow

$$\Phi : X^\phi \times \mathbb{R}^+ \to X^\phi$$

such that

$$\Phi\left((x,t), t'\right) = \left(\sigma^{n(x,t+t')}, t + t' - s(x, t+t')\right)$$

with $n(\cdot,\cdot)$ and $s(\cdot,\cdot)$ as before.

10.2. Bowen–Walters' distance

For $\phi = 1$ there exists a natural metric compatible with the quotient topology on X^1, which was first introduced in [31]. This definition can be readily adapted for the general case.

Denote $(x, t\phi(x))$ by x_t, and consider the t-horizontal sections

$$X_t := \left\{\left(x, t\phi(x)\right) \in X^\phi\colon x \in X\right\}, \quad t \in [0,1).$$

Supply the t-horizontal section with the distance

$$\rho_t(x_t, y_t) = (1-t)d_X(x, y) + t d_X(\sigma x, \sigma y).$$

A path p between x_t and $y_{t'}$ is a finite sequence

$$p = \{x_t = x_{t_0}^{(0)}, x_{t_1}^{(1)}, \dots, x_{t_{n-1}}^{(n-1)}, x_{t_n}^{(n)} = y_{t'}\},$$

such that for each $0 \leqslant i < n$, either $x^{(i+1)} \in \{x^{(i)}, \sigma x^{(i)}\}$, or $t_i = t_{i+1}$. The length of the path p is given by $|p| \equiv \sum_{i=0}^{n-1} |\{x_{t_i}^{(i)}, x_{t_{i+1}}^{(i+1)}\}|$, with

$$|\{x_{t_i}^{(i)}, x_{t_{i+1}}^{(i+1)}\}| = \begin{cases} 1 - t_i + t_{i+1} & \text{if } \sigma x^{(i)} = x^{(i+1)}, \\ \rho_{t_i}(x_{t_i}^{(i)}, x_{t_i}^{(i+1)}) + |t_{i+1} - t_i| & \text{otherwise.} \end{cases}$$

Finally, the distance in the suspended space is given by

$$d_{X^\phi}(x_t, y_{t'}) = \inf\{|p| \colon p \in [x_t \to y_{t'}]\},$$

where $[x_t \to y_{t'}]$ denotes the set of all paths from x_t to $y_{t'}$.

10.3. Spectrum of dimensions

In the computation of the Poincaré spectrum for Φ, we will use covers of X^ϕ by suspended open sets we call rectangles.

Given $\varepsilon > 0$ and $x_t \in X^\phi$, define the t-horizontal open ball of radius ε, $\mathcal{S}(x_t, \varepsilon) = \{y_t \in X_t \colon \rho_t(x_t, y_t) < \varepsilon\}$. Let us remind that $y_t := (y, t\phi(y))$. The rectangle with base $\mathcal{S}(x_t, \varepsilon)$ and of height $\delta > 0$ is the set

$$R(x_t, \varepsilon, \delta) = \bigcup_{y_t \in \mathcal{S}(x_t, \varepsilon)} \bigcup_{0<s<\delta} \Phi(y_s, \phi(y)s).$$

These sets are of course open in X^ϕ.

10.3.1. The Poincaré recurrence

The Poincaré recurrence $\tau_\Phi(B)$, for a set $B \subset X^\phi$, is the minimal time that an orbit starting in B spends in $X^\phi \setminus B$ before it reenters B. In the particular case of a rectangle $R(x_t, \varepsilon, \delta)$, the Poincaré recurrence can be computed as follows. For each $y_t \in \mathcal{S}(x_t, \varepsilon)$ let

$$\tau_\Phi(y_t, R(x_t, \varepsilon, \delta)) = \inf\{s \geqslant \phi(y)\delta \colon \Phi(y_t, s) \in R(x_t, \varepsilon, \delta)\},$$

then

$$\tau_\Phi(R(x_t, \varepsilon, \delta)) = \inf\{\tau_\Phi(y_t, R(x_t, \varepsilon, \delta)) \colon y_t \in R(x_t, \varepsilon, \delta)\}.$$

10.3.2. The spectrum

Let $\mathcal{R}$ be a finite cover of X^ϕ by rectangles $R_i \equiv R(x_{t_i}^{(i)}, \varepsilon_i, \delta_i)$, $\alpha, q \in \mathbb{R}$, and $Z \subset X^\phi$. Define

$$\mathcal{M}(Z, \alpha, q, \mathcal{R}) = \sum_{R_i \in \mathcal{R}} \exp\bigl(-q\tau_\Phi(R_i)\bigr)(\operatorname{diam} R_i)^\alpha.$$

Monotonicity implies that the limit

$$\mathcal{M}(Z, \alpha, q) \equiv \lim_{\varepsilon \to 0}\Bigl(\inf_{\operatorname{diam}(\mathcal{R}) \leqslant \varepsilon} \mathcal{M}(Z, \alpha, q, \mathcal{R})\Bigr)$$

exists. Here, by diam $(\mathcal{R})$ we mean the maximum of the diameters of rectangles in $\mathcal{R}$, and the infimum is taken over all finite covers of X^ϕ by rectangles satisfying the condition diam $(\mathcal{R}) \leqslant \varepsilon$. Note that for each fixed $q \in \mathbb{R}$, $\mathcal{M}(\alpha, q) = \infty$ for α small enough. We can now define the spectrum for Poincaré recurrences for $Z \subset X^\phi$ as

$$\alpha(Z, q) \equiv \sup\bigl\{\alpha \in \mathbb{R}\colon \mathcal{M}(\alpha, q) = \infty\bigr\}.$$

10.3.3. Main results

THEOREM 10.1. *For $\alpha \geqslant 1$ and $q \geqslant 0$, the spectrum for Poincaré recurrences $\alpha^* \equiv \alpha(X^\phi, q)$ satisfies the equation*

$$P_X\bigl((1-\alpha^*)u - q\phi|\sigma\bigr) = 0.$$

The validity of the equation follows from two inequalities.

CLAIM 10.1. *If α and $q \in \mathbb{R}$ are such that $P_X((1-\alpha)u - q\phi|\sigma) < 0$, then $\alpha \geqslant \alpha^* \equiv \alpha(X^\phi, q)$.*

CLAIM 10.2. *If $\alpha \geqslant 1$ and $q \geqslant 0$ are such that $P_X((1-\alpha)u - q\phi|\sigma) > 0$, then $\alpha \leqslant \alpha^* \equiv \alpha(X^\phi, q)$.*

The first inequality can be proven by exhibiting a sequences $\{\mathcal{R}_n\}_{n=1}^\infty$ of covers by rectangles such that $\lim_{n\to\infty} \mathcal{M}(X^\phi, \alpha, q, \mathcal{R}_n) < \infty$, for all α satisfying the inequality $P_X((1-\alpha)u - q\phi|\sigma) < 0$. The second statement requires a variational reasoning. In both cases we will need the following results.

LEMMA 10.1. *Let $x_t, y_{t'} \in X^\phi$, with $\zeta^1(x) = \zeta^1(y)$, be such that $\rho_s(x_s, y_s) + 2|t - t'| \leqslant 1$, where $s = \min(t, t')$. Then $d_{X^\phi}(x_t, y_{t'}) = \rho_s(x_s, y_s) + |t' - t|$.*

PROOF. We will say that the path

$$p = \{z_{t_0}^{(0)}, z_{t_1}^{(1)}, \dots, z_{t_{m-1}}^{(m-1)}, z_{t_m}^{(m)}\} \in [x_t \to y_{t'}]$$

is traversing at the point $z_{t_k}^{(k)}$ if $z^{(k+1)} = \sigma z^{(k)}$. On the contrary, the path p is non-traversing if $z^{(i)} = z^{(i+1)}$ whenever $t_i \neq t_{i+1}$.

Traversing paths. Note that for a non-traversing path $q \in [z_{t_0}^{(0)} \to z_{t_m}^{(m)}]$ we have that

$$|q| \geqslant \sum_{i=1}^{m} \left(\rho_{s_i}\left(z_{s_i}^{(i)}, z_{s_i}^{(i-1)}\right) + |t_{i-1} - t_i|\right) \geqslant \sum_{i=1}^{m} |t_i - t_{i-1}| \geqslant |t_m - t_0|.$$

Now suppose that

$$p = \{z_{t_0}^{(0)}, z_{t_1}^{(1)}, \dots, z_{t_{m-1}}^{(m-1)}, z_{t_m}^{(m)}\} \in \left[z_{t_0}^{(0)} := x_t \to z_{t_m}^{(m)} := y_{t'}\right]$$

traverses only at $z_{t_k}^{(k)}$, so that both $p_0 := \{z_{t_0}^{(0)}, \dots, z_{t_k}^{(k)}\}$ and $p_1 := \{z_{t_{k+1}}^{(k+1)}, \dots, z_{t_m}^{(m)}\}$ are non-traversing. In this case

$$|p| = |p_0| + |p_1| + 1 - t_k + t_{k+1} \geqslant |t_0 - t_k| + |t_{k+1} - t_m| + 1 - t_k + t_{k+1}.$$

Taking into account that $t_{k+1} - t_k \leqslant |t_k - t_0| + |t_0 - t_m| + |t_{k+1} - t_m|$ and the hypothesis $\rho_s(x_s, y_s) + 2|t - t'| \leqslant 1$, we readily obtain

$$|p| \geqslant 1 - |t_0 - t_m| \geqslant \rho_s(x_s, y_s) + |t - t'|.$$

If $p \in [z_{t_0}^{(0)} := x_t \to z_{t_m}^{(m)} := y_{t'}]$ traverses at ℓ different points, then we can write

$$p = \{z_{t_0}^{(0)} := z_{t_{k_0}}^{(k_0)}, \dots, z_{t_{k_1}}^{(k_1)}, \dots, z_{t_{k_2}}^{(k_2)}, \dots, z_{t_{k_\ell}}^{(k_\ell)} := z_{t_m}^{(m)}\},$$

where for each $j = 0, 1, \dots, \ell - 1$ we have that the section $p_j := \{z_{t_{k_j}}^{(k_j)}, \dots, z_{t_{k_{j+1}}}^{(k_{j+1})}\}$ traverses only at one point, and $t_{k_j} = t_0$. For this it is enough to add, if necessary, intermediate points to the original path. Hence, according to the inequality we just obtained,

$$|p| = \sum_{j=1}^{\ell-1} |p_j| \geqslant \ell - \sum_{j=1}^{\ell-1} |t_{k_{j+1}} - t_{k_j}| = \ell - |t_m - t_0|,$$

from which we derive $|p| \geqslant \rho_s(x_s, y_s) + |t - t'|$.

Therefore, for each traversing path $p \in [x_t \to y_{t'}]$ we have $|p| \geqslant \rho_s(x_s, y_s) + |t - t'|$ as far as the hypothesis $\rho_s(x_s, y_s) + 2|t - t'| \leqslant 1$ holds.

Non-traversing paths. First consider

$$p = \{z_{t_0}^{(0)}, z_{t_1}^{(1)}, \ldots, z_{t_{m-1}}^{(m-1)}, z_{t_m}^{(m)}\} \in [x_t := z_{t_0}^{(0)} \to y_{t'} := z_{t_m}^{(m)}]$$

such that $\zeta^1(z^{(i)}) = \zeta^1(z^{(0)})$ for $i = 0, 1, \ldots, m$, that is to say, the X-projection of p is completely contained in the 1-cylinder $\zeta^1(x)$. For p of this kind we have $d_X(\sigma z^{(i)}, \sigma z^{(i+1)}) \geqslant d_X(z^{(i)}, z^{(i+1)})$ for $i = 0, 1, \ldots, m-1$. In this case, for $s \leqslant t$ we have

$$\begin{aligned}\rho_t(z_t^{(i)}, z_t^{(i+1)}) &\equiv (1-t)d_X(z^{(i)}, z^{(i+1)}) + td_X(\sigma z^{(i)}, \sigma z^{(i+1)}) \\ &= (1-s)d_X(z^{(i)}, z^{(i+1)}) + sd_X(\sigma z^{(i)}, \sigma z^{(i+1)}) \\ &\quad + (t-s)(d_X(\sigma z^{(i)}, \sigma z^{(i+1)}) - d_X(z^{(i)}, z^{(i+1)})) \\ &\geqslant \rho_x(z_s^{(i)}, z_s^{(i+1)})\end{aligned}$$

for each $i = 0, 1, \ldots, m-1$. Let $s' = \min_i t_i$ and $s = \min(t_0, t_m)$, then (see illustration in Figure 10.1), then

$$\begin{aligned}|p| &= \sum_{i=0}^{m-1} (\rho_{t_i}(z_{t_i}^{(i)}, z_{t_i}^{(i+1)}) + |t_{i+1} - t_i|) \\ &\geqslant \sum_{i=0}^{m-1} \rho_{s'}(z_{s'}^{(i)}, z_{s'}^{(i+1)}) + \sum_{i=0}^{m-1} |t_{i+1} - t_i| \\ &\geqslant \rho_{s'}(z_{s'}^{(0)}, z_{s'}^{(m)}) + |t_m - t_0| + 2(s - s') \\ &= \rho_s(x_s^{(0)}, x_s^{(m)}) + |t_0 + t_m| + (s - s')(2 - \delta(y^{(0)}, y^{(m)})) \\ &\geqslant \rho_s(x_s^{(0)}, x_s^{(m)}) + |t_0 - t_m|.\end{aligned}$$

Therefore, each non-traversing path $p \in [x_t \to y_t']$ whose X projection belongs to the a single 1-cylinder satisfies the inequality $|p| \geqslant \rho_s(x_s, y_s) + |t - t'|$.

For a general non-traversing path, each one of its segments with the X-projection entirely contained in a single 1-cylinder can be replace by a minimal path with the same end points. In this way the length of a general non-traversing path $p \in [x_t \to y_{t'}]$ can be bounded from below by the length of a concatenation

$$q = \{z_{t_0}^{(0)}\}\alpha_0\beta_{0,1}\cdots\beta_{m-1,m}\alpha_m\{z_{t_{m+1}}^{(2m+1)}\} \in [x_t \to y_{t'}],$$

where for $j = 0, 1, \ldots, m$, the path $\alpha_j = \{z_{s_j}^{(2j)}, z_{s_j}^{(2j+1)}\}$, with $s_j = \min(t_j, t_{j+1})$ and $\zeta^1(z^{(2j)}) = \zeta^1(z^{(2j+1)})$, replaces a section of p whose X-projections is completely included a 1-cylinder; for $j = 0, 1, \ldots, m-1$, the path $\beta_{j,j+1} = \{z_{t_{j+1}}^{(2j+1)}, z_{t_{j+1}}^{(2j+2)}\}$, with $\zeta^1(z^{(2j+1)}) \neq \zeta^1(z^{(2j+2)})$, corresponds to the section

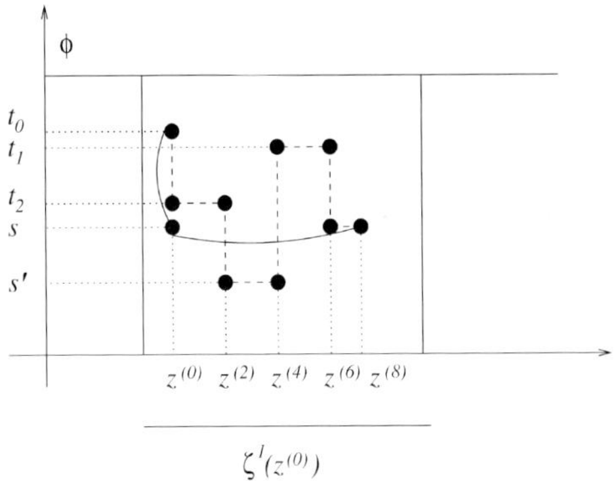

Figure 10.1. The broken line represents a non-traversing path whose X-projection is entirely contained in the 1-cylinder $\zeta^1(z^{(0)})$. The continuous curve represents the corresponding path having a shorter length.

of p whose X-projection joints consecutive 1-cylinders (see illustration in Figure 10.2). Since

$$d_X(z, z') = 1 \quad \text{whenever } \zeta^1(z) \neq \zeta^1(z'),$$

then $|\beta_{j,j+1}| = 1 - t_j + t_j d_X(\sigma z^{(2j-1)}, \sigma z^{(2j)})$ for $j = 0, 1, \ldots, m-1$.

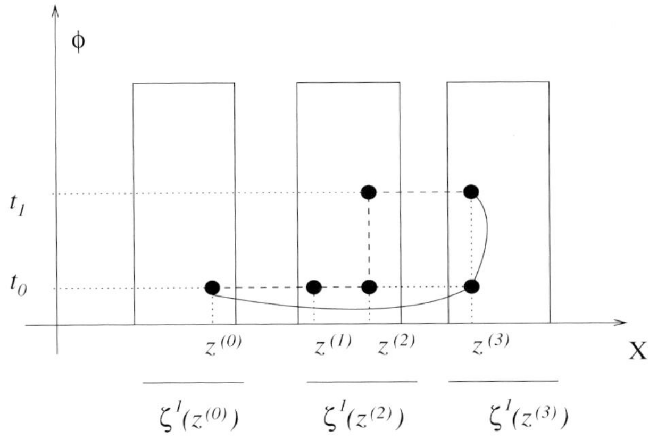

Figure 10.2. The broken line represents a non-traversing path whose X-projection intersects three 1-cylinders. The continuous curve represents the corresponding path intersecting only the first and last 1-cylinders, and having a shorter length.

The number of different 1-cylinders intersecting the X-projection of the path q may be reduced as follows. For

$$\beta_{0,1}\alpha_1\beta_{1,2} := \{z_{t_1}^{(1)}, z_{t_1}^{(2)}, z_{s_1}^{(2)}, z_{s_1}^{(3)}, z_{t_2}^{(3)}, z_{t_2}^{(4)}\},$$

we have that

$$\begin{aligned}|\beta_{0,1}\alpha_1\beta_{1,2}| &= \rho_{s_1}\big(z_{s_1}^{(2)}, z_{s_1}^{(3)}\big) + |t_1 - t_2| + 1 - t_1 + 1 - t_2 \\ &\quad + t_1 d_X\big(\sigma z^{(1)}, \sigma z^{(2)}\big) + t_2 d_X\big(\sigma z^{(3)}, \sigma z^{(4)}\big) \\ &\geqslant \rho_{s_1}\big(z_{s_1}^{(1)}, z_{s_1}^{(4)}\big) + |t_1 - t_2| + 1 - t_1 + 1 - t_2 \\ &\quad + t_1 d_X\big(\sigma z^{(1)}, \sigma z^{(2)}\big) + t_2 d_X\big(\sigma z^{(3)}, \sigma z^{(4)}\big) \\ &\quad + s_1\big(d_X\big(\sigma z^{(2)}, \sigma z^{(3)}\big) - d_X\big(\sigma z^{(1)}, \sigma z^{(4)}\big)\big) \\ &\quad + (1 - s_1)\big(d_X\big(z^{(2)}, \sigma z^{(3)}\big) - d_X\big(\sigma z^{(1)}, \sigma z^{(4)}\big)\big).\end{aligned}$$

Taking into account that $t_1, t_2 \geqslant s_1$, and using the triangle inequality, we obtain

$$\begin{aligned}|\beta_{0,1}\alpha_1\beta_{1,2}| &\geqslant \rho_{s_1}\big(z_{s_1}^{(1)}, z_{s_1}^{(4)}\big) + |t_1 - t_2| + 1 - \max(t_1, t_2) \\ &\quad + s_1\big(d_X\big(\sigma z^{(2)}, \sigma z^{(3)}\big) + d_X\big(\sigma z^{(1)}, \sigma z^{(2)}\big) \\ &\quad + d_X\big(\sigma z^{(3)}, \sigma z^{(4)}\big) - d_X\big(\sigma z^{(1)}, \sigma z^{(4)}\big)\big) \\ &\quad + (1 - s_1)\big(1 + d_X\big(z^{(2)}, \sigma z^{(3)}\big) - d_X\big(\sigma z^{(1)}, \sigma z^{(4)}\big)\big) \\ &\geqslant \rho_{s_1}\big(z_{s_1}^{(1)}, z_{s_1}^{(4)}\big) + |t_1 - t_2| + 1 - \max(t_1, t_2) \\ &\geqslant \big|\{z_{t_1}^{(1)}, z_{s_1}^{(1)}, z_{s_1}^{(4)}, z_{t_2}^{(4)}\}\big|.\end{aligned}$$

Hence, $|q| \geqslant |q'|$ where

$$q' := \begin{cases} \{z_{t_0}^{(0)}\}\alpha_0\{z_{t_1}^{(1)}, z_{t_1}^{(4)}, z_{t_2}^{(4)}\}\alpha_2 \cdots \alpha_m\{z_{t_{m+1}}^{(2m+1)}\} & \text{if } t_1 \leqslant t_2, \\ \{z_{t_0}^{(0)}\}\alpha_0\{z_{t_1}^{(1)}, z_{t_2}^{(1)}, z_{t_2}^{(4)}\}\alpha_2 \cdots \alpha_m\{z_{t_{m+1}}^{(2m+1)}\} & \text{if } t_1 > t_2. \end{cases}$$

This can be further reduce, so that $|q'| \geqslant |q''|$ with

$$q'' := \begin{cases} \{z_{t_0}^{(0)}\}\alpha_0\{z_{t_1}^{(1)}, z_{t_1}^{(4)}, z_{s_1'}^{(4)}, z_{s_1'}^{(5)}, z_{t_3}^{(5)}\}\beta_{2,3} \cdots \alpha_m\{z_{t_{m+1}}^{(2m+1)}\} & \text{if } t_1 \leqslant t_2, \\ \{z_{t_0}^{(0)}, z_{s_0'}^{(0)}, z_{s_0'}^{(1)}, z_{t_2}^{(1)}, z_{t_2}^{(4)}\}\alpha_2 \cdots \alpha_m\{z_{t_{m+1}}^{(2m+1)}\} & \text{if } t_1 > t_2. \end{cases}$$

In this way we obtain a path $q'' \in [x_t \to y_{t'}]$ whose X-projection intersects one less 1-cylinder than p, and such that $|p| \geqslant |q''|$. By repeating this reduction, we finally obtain

$$|p| \geqslant \big|\{z_{t_0}^{(0)}, z_{s_\beta}^{(0)}, z_{s_\beta}^{(1)}, z_{s_\beta}^{(2m)}, z_{s_\beta}^{(2m+1)}, z_{t_{m+1}}^{(2m+1)}\}\big|$$

where $s_\beta = \min\{t_i\colon i = 1, 2, \ldots, m\}$. Since the X-projection of the path

$$\left\{z_{t_0}^{(0)}, z_{s_\beta}^{(0)}, z_{s_\beta}^{(1)}, z_{s_\beta}^{(2m)}, z_{s_\beta}^{(2m+1)}, z_{t_{m+1}}^{(2m+1)}\right\} \in [x_t \to y_{t'}]$$

is completely contained in $\zeta^1(x)$, then $|p| \geqslant \rho_s(x_s, y_s) + |t - t'|$.

To conclude the proof, note that the path $\{x_t, x_s, y_s, y_{t'}\}$ reaches the minimum length. □

REMARK 10.1. Note that for arbitrary x_t, y_t',

$$d_{X^\phi}(x_t, y_{t'}) \leqslant \rho_s(x_s, y_s) + |t - t'|,$$

since $p = \{x_t, x_s, y_s, y_{t'}\}$ has length $\rho_s(x_s, y_s) + |t - t'|$. Therefore,

$$\operatorname{diam} R(x_t, \varepsilon, \delta) \leqslant \delta + \max_{0<\eta<\delta} \operatorname{diam} \mathcal{S}(x_{s+\eta}, \varepsilon).$$

If $s + \delta < 1$, then we have that

$$\begin{aligned}\max_{0<\eta<\delta} \operatorname{diam} \mathcal{S}(x_{s+\eta}, \varepsilon) \leqslant \delta & \max_{z,z' \in \mathcal{S}(x_s,\varepsilon)} \bigl(d_X(\sigma z, \sigma z') - d_X(z, z')\bigr) \\ & + \operatorname{diam} \mathcal{S}(x_s, \varepsilon) \leqslant \varepsilon + \delta.\end{aligned}$$

If $\delta + s > 1$ then

$$\begin{aligned}\max_{0<\eta<\delta} \operatorname{diam} \mathcal{S}(x_{s+\eta}, \varepsilon) & \leqslant \max_{z,z' \in \mathcal{S}(x_s,\varepsilon)} \rho_\delta\bigl((\sigma z)_\delta, (\sigma z')_\delta\bigr) \\ & \leqslant e^{\max u} \operatorname{diam} \mathcal{S}(x_s, \varepsilon) \leqslant e^{\max u} \varepsilon.\end{aligned}$$

In both cases $\operatorname{diam} R(x_t, \varepsilon, \delta) \leqslant e^{\max u}\varepsilon + 2\delta$.

For each rectangle $R(x_s, \varepsilon, \delta)$ let $R_X(x_s, \varepsilon) := \{y \in X\colon\ y_s \in \mathcal{S}(x_s, \varepsilon)\}$ be its X-projection, which does not depend on δ. Furthermore, as we prove below, it is a cylinder sets.

PROPOSITION 10.1. *For each $x_s \in X^\phi$ and $\varepsilon > 0$ there exists $m(x_s, \varepsilon) \in \mathbb{N}$ such that $R_X(x_s, \varepsilon) = \zeta^{m(x_s,\varepsilon)}(x)$.*

PROOF. For every $y \in X \setminus \{x\}$, the values of the distances $\rho_s(x_s, y_s)$ belong to the countable set

$$D_{x,s} := \left\{d_{x,s,n} := (1 - s)e^{-u(\zeta^n(x))} + se^{-u(\zeta^{n-1}(\sigma x))}\colon\ n \in \mathbb{N}\right\},$$

with no accumulation points other than zero. Furthermore, since $u > 0$, then $d_{x,s,n} = d_{x,s,m}$ if and only if $m = n$. Clearly $d_{x,s,n}$ decreases with n, so that the elements in $D_{x,s}$ form a sequence $d_{x,s,0} > d_{x,s,1} > \cdots$ converging to zero. Since $y \in R_X(x_s, \varepsilon) \Longleftrightarrow \rho_s(x_s, y_s) = d_{x,s,m} < \varepsilon$ with

$$m = \max\left\{n \in \mathbb{N}\colon\ \zeta^n(x) = \zeta^n(y)\right\},$$

then the result follows with

$$m(x_s, \varepsilon) = \min\{n \in \mathbb{N}\colon d_{x,s,n} < \varepsilon\}. \qquad \square$$

LEMMA 10.2. *There exists a constant* $\bar{a} \geqslant 0$ *such that for any* $x_s \in X^\phi$ *and* $0 < \delta, \varepsilon < 1$,

$$\max_{z \in R_X(x_s,\varepsilon)} \sum_{j=0}^{\tau} \phi(\sigma^j z) - \bar{a} \leqslant \tau_\Phi\bigl(R(x_s, \varepsilon, \delta)\bigr) \leqslant \max_{z \in R_X(x_s,\varepsilon)} \sum_{j=0}^{\tau} \phi(\sigma^j z) + \bar{a}$$

where $\tau := \tau_\sigma(R_X(x_s, \varepsilon))$.

PROOF. Let $\tau = \tau_\sigma(R_X(x_s, \varepsilon))$. For each $z \in R_X(x_s, \varepsilon) \cap \sigma^{-\tau}(R_X(x_s, \varepsilon))$, and each $\eta \in (0, \min(\delta, 1 - s))$, if

$$\begin{aligned} & s\bigl(\phi\bigl(\sigma^\tau(z)\bigr) - \phi(z)\bigr) - \eta\phi(z) \\ & \quad < t - \sum_{j=0}^{\tau-1} \phi\bigl(\sigma^j(z)\bigr) < (s + \eta)\bigl(\phi\bigl(\sigma^\tau(z)\bigr) - \phi(z)\bigr), \end{aligned}$$

then both $z_{s+\eta}$ and $\Phi(z_{s+\eta}, t)$ belong to $R(x_s, \varepsilon, \delta)$. Therefore

$$\tau_\Phi\bigl(R(x_s, \varepsilon, \delta)\bigr) \leqslant \max_{z \in R_X(x_s,\varepsilon)} \sum_{j=0}^{\tau} \phi(\sigma^j z) + 2 \max \phi.$$

In this way we prove one of the inequalities. Now, for each $z \in R_X(x_s, \varepsilon)$ let $\tau(z) := \min\{k \in \mathbb{N}\colon \sigma^k(z) \in R_X(x_s, \varepsilon)\}$. Since $\tau = \min_{z \in R_X(x_s,\varepsilon)} \tau(z)$, then

$$\begin{aligned} \tau_\Phi\bigl(R(x_s, \varepsilon, \delta)\bigr) & \geqslant -(s + \delta) \max \phi + \min_{z \in R_X(x_s,\varepsilon)} \sum_{j=0}^{\tau(z)-1} \phi(\sigma^j z) \\ & \geqslant -(s + \delta) \max \phi + \min_{z \in R_X(x_s,\varepsilon)} \sum_{j=0}^{\tau-1} \phi(\sigma^j z), \end{aligned}$$

regardless the value of $\delta \in (0, 1)$. Furthermore, since ϕ is a Hölder continuous function, then there exists a constant $\theta > 0$ such that $|\phi(z) - \phi(y)| \leqslant \exp(-\theta u(\zeta^n(z)))$, with $n = \max\{k \in \mathbb{N}\colon \zeta^k(y) = \zeta^k(z)\}$. Therefore

$$\begin{aligned} & \tau_\Phi\bigl(R(x_s, \varepsilon, \delta)\bigr) \\ & \quad \geqslant \max_{z \in R_X(x_s,\varepsilon)} \sum_{j=0}^{\tau} \phi(\sigma^j z) \\ & \qquad - \left((s + \varepsilon + 1) \max \phi + \frac{1}{1 - \exp(-\theta \min \phi)}\right) \end{aligned}$$

with $\max\phi = \max_{z\in X}\phi(z)$ and similarly for $\min\phi$. Finally, the result follows with $\bar{a} = 3\max\phi + (1 - \exp(-\theta\min\phi))^{-1}$. □

10.3.4. Proof of Claim 10.1

For each $n \geqslant 2$ let $X_n = \{x^{(1)}, x^{(2)}, \ldots, x^{p(n)}\} \subset X$ be such that $\zeta^n = \{\zeta^n(x^{(i)})\colon x^{(i)} \in X_n\}$. For each $i \in \{1, 2, \ldots, p(n)\}$ define $s(i, 0) = 0$, and recursively

$$\begin{aligned}\varepsilon_{i,j} &= \sup\{\varepsilon > 0\colon R_X\big(x^{(i)}_{s(i,j)}, \varepsilon\big) = \zeta^n\big(x^{(i)}\big)\},\\ s(i, j+1) &= s(i, j) + \varepsilon_{i,j}\big(1 - e^{-n\max u}\big),\end{aligned}$$

for every $j \geqslant 1$. Note that $\varepsilon_{i,0} = \varepsilon_{n,x}$, and both $s(i, j)$ and $\varepsilon_{i,j}$ increase with j. Finally, for each $i \in \{1, 2, \ldots, p(n)\}$ let $n(i) = \max\{j \geqslant 0\colon s(i, j) \leqslant 1\}$. The collection

$$\mathcal{R}_n = \big\{R_{i,j} := R\big(x^{(i)}_{s(i,j)}, \varepsilon_{i,j}, \varepsilon_{i,j}\big)\colon x^{(i)} \in X_n,\ j \in \{0, 1, \ldots, n(i)\}\big\},$$

is a cover of X^ϕ by (squared) rectangles. We use this particular cover to compute an upper bound for the spectra. Applying Remark 10.1, Lemma 10.2, and taking into account the definition of $R_{i,j}$, we obtain

$$\begin{aligned}\mathcal{M}\big(X^\phi, \alpha, q, \mathcal{R}_n\big) &= \sum_{i=1}^{p(n)}\sum_{j=0}^{n(i)} \exp\big(-q\tau_\phi(R_{i,j})\big)\operatorname{diam} R_{i,j}^\alpha\\ &\leqslant e^{q\bar{a}}\sum_{i=1}^{p(n)}\sum_{j=0}^{n(i)} e^{-q\phi(\zeta^{\tau_i}(x^{(i)})}\big(\big(2 + e^{\max u}\big)\varepsilon_{i,j}\big)^\alpha\\ &\leqslant e^{q\bar{a}}\big(2 + e^{\max u}\big)^\alpha\sum_{i=1}^{p(n)}\sum_{j=0}^{n(i)} e^{-q\phi(\zeta^{\tau_i}(x^{(i)}))}\varepsilon_{i,j}^\alpha\end{aligned}$$

with $\tau_i = \tau_\sigma(\zeta^n(x^{(i)}))$. Furthermore, since

$$\varepsilon_{i,j} < e^{-u(\zeta^{n-1}(\sigma x^{(i)}))} < e^{-u(\zeta^n(x^{(i)}))+2\max u},$$

we have

$$\mathcal{M}\big(X^\phi, \alpha, q, \mathcal{R}_n\big) \leqslant e^{q\bar{a}}\big(3e^{3\max u}\big)^\alpha\sum_{i=1}^{p(n)}\big(n(i) + 1\big)e^{-q\phi(\zeta^{\tau_i}(x^{(i)}))-\alpha u(\zeta^n(x^{(i)}))}.$$

On the other hand, since $e^{-u(\zeta^n(x^{(i)}))} \leqslant \varepsilon_{i,j}$ and

$$1 \geqslant s\big(i, n(i)\big) = \big(1 - \exp(-n\max u)\big)\left(\sum_{j=0}^{n(i)-1}\varepsilon_{i,j}\right)$$

we deduce the inequality $n(i) \leqslant \exp(u(\zeta^n(x^{(i)})))(1 - \exp(-n \max u))^{-1}$, and from this we obtain

$$\begin{aligned}
&\mathcal{M}(X^\phi, \alpha, q, \mathcal{R}_n) \\
&\quad \leqslant C_0 \sum_{i=1}^{p(n)} \left(\frac{e^{u(\zeta^n(x^{(i)})}}{1 - e^{-n \max u}} + 1 \right) e^{-q\phi(\zeta^{\tau_i}(x^{(i)})) - \alpha u(\zeta^n(x^{(i)}))} \\
&\quad \leqslant \frac{C_0(1 + e^{n \min u})}{1 - e^{-n \max u}} \sum_{i=1}^{p(n)} e^{-q\phi(\zeta^{\tau_i}(x^{(i)})) + (1-\alpha)u(\zeta^n(x^{(i)}))},
\end{aligned}$$

with $C_0 := 3^\alpha \exp(q\bar{a} + 3\alpha \max u)$. Next, define $P_{n,k} = \{1 \leqslant i \leqslant p(n)\colon \tau_i = k\}$. Since (X, σ) is specified, then $P_{n,k} = \emptyset$ for all $k \geqslant n + n_0$. We have,

$$\mathcal{M}(X, \alpha, q, \mathcal{R}_n) \leqslant C_1 e^{n_0 \max u} \sum_{k=1}^{n+n_0} \sum_{i \in P_{n,k}} e^{-q\phi(\zeta^k(x^{(i)})) + (1-\alpha)u(\zeta^k(x^{(i)}))},$$

with $C_1 := C_0(1 + e^{-n \min u})(1 - e^{-n \max u})^{-1}$. Now, $\alpha, q \in \mathbb{R}$ such that $\delta + P_X(-q\phi + (1-\alpha)u|\sigma)) < 0$ there exists $\delta > 0$ and $k_\delta \in \mathbb{N}$ such that

$$\sum_{c \in \zeta^k} e^{-q\phi(c) + (1-\alpha)u(c)} \leqslant e^{-\delta/2},$$

for all $k \geqslant k_\delta$. Then, since $\{\zeta^k(x^{(i)})\colon i \in P_{n,k}\} \subset \zeta^k$, we obtain

$$\begin{aligned}
\mathcal{M}(X, \alpha, q, \mathcal{R}_n) &\leqslant C_1 e^{n_0 \max u} \sum_{k=1}^{\infty} \sum_{c \in \zeta^k} e^{-q\phi(c) + (1-\alpha)u(c)} \\
&\leqslant C_1 e^{n_0 \max u} \left(\Sigma_{k_\delta} + \frac{e^{-k_\delta \times \delta/2}}{1 - e^{-\delta/2}} \right) < \infty,
\end{aligned}$$

with

$$\Sigma_{k_\delta} := \sum_{k=1}^{k+\delta} \sum_{c \in \zeta^k} e^{-q\phi(c) + (1-\alpha)u(c)}.$$

Now, $C_1 = C_1(n)$ tends to C_0 as n goes to infinity, therefore

$$\begin{aligned}
\mathcal{M}(X^\phi, \alpha, q) &\leqslant \lim_{n \to \infty} \mathcal{M}(X^\phi, \alpha, q, \mathcal{R}_n) \\
&\leqslant C_0 e^{n_0 \max u} \left(\Sigma_{k_\delta} + \frac{e^{-k_\delta \times \delta/2}}{1 - e^{-\delta/2}} \right) < \infty
\end{aligned}$$

as far as $P_X(-q\phi + (1-\alpha)u|\sigma) < -\delta$. With this the proof is done. □

10.3.5. Proof of Claim 10.2

Vertical dimension. Given a cover $\mathcal{R} = \{R(x_s, \varepsilon, \delta)\}$ of X^ϕ by rectangles, to each rectangles $R(x_s, \varepsilon, \delta) \in \mathcal{R}$ we associate the integer $n(x_s, \varepsilon, \delta) := \lceil \delta \times e^{u(R_X(x_s, \varepsilon))} \rceil$.

For each $R(x_s, \varepsilon, \delta) \in \mathcal{R}$ such that $s + \delta < 1$ and $\varepsilon + 2\delta \leqslant 1$, Lemma 10.1 implies that

$$\begin{aligned} \operatorname{diam} R(x_s, \varepsilon, \delta) &= \operatorname{diam} \mathcal{S}(x_s, \varepsilon) + \delta \\ &= \varepsilon + \delta \\ &\geqslant \varepsilon + e^{-u(R_X(x_s, \varepsilon))}\big(n(x_s, \varepsilon, \delta) - 1\big). \end{aligned}$$

Taking into account that $\varepsilon \geqslant \exp(-u(R_X(x_s, \varepsilon)))$, we obtain

$$\operatorname{diam} R(x_s, \varepsilon, \delta) \geqslant e^{-u(R_X(x_s, \varepsilon))} n(x_s, \varepsilon, \delta).$$

Therefore, for $\alpha \geqslant 1$ we have

$$\big\{\operatorname{diam} R(x_s, \varepsilon, \delta)\big\}^\alpha \geqslant e^{-\alpha u(R_X(x_s, \varepsilon))} n(x_s, \varepsilon, \delta).$$

Now, since $n(x_s, \varepsilon, \delta) \geqslant \delta e^{u(R_X(x_s, \varepsilon))}$, we obtain that

$$\big\{\operatorname{diam} R(x_s, \varepsilon, \delta)\big\}^\alpha \geqslant \delta e^{-(\alpha+1) u(R_X(x_s, \varepsilon))},$$

for each $R(x_s, \varepsilon, \delta) \in \mathcal{R}$ with $s + \delta < 1$ and $\varepsilon + 2\delta \leqslant 1$.

Horizontal dimension. Fix $N \in \mathbb{N}$ and for each $0 \leqslant j < N$ let

$$\mathcal{S}_{N,j} := \big\{x_t \in X^\phi \colon\ x \in X \text{ and } j/N < t < (j+1)/N\big\}$$

be the jth horizontal slice. The jth slice is properly covered by $\mathcal{R}$ if the collection

$$\mathcal{R}_{N,j} := \big\{R(x_s, \varepsilon, \delta) \in \mathcal{R} \colon\ s < j/N < (j+1)/N < s + \delta\big\}$$

is a cover for $\mathcal{S}_{N,j}$. If $\mathcal{S}_{N,j}$ is properly covered by $\mathcal{R}$, then the projection

$$\mathcal{C}_{N,j} := \big\{R_X(x_s, \varepsilon) \colon\ R(x_s, \varepsilon, \delta) \in \mathcal{R}_{N,j}\big\}$$

is a cover for X.

The number of covers $\mathcal{R}_{N,j}$ containing a given rectangle $R(x_s, \varepsilon, \delta) \in \mathcal{R}$ is bounded. Indeed, given $R(x_s, \varepsilon, \delta) \in \mathcal{R}$ we have

$$\#\big\{0 \leqslant j < N \colon\ R(x_s, \varepsilon, \delta) \in \mathcal{R}_{N,j}\big\} \leqslant \delta \times N.$$

For N sufficiently large, the majority of the horizontal slices are properly covered by rectangles in $\mathcal{R}$. Indeed, if $\mathcal{S}_{N,j}$ is not properly covered, then it intersects at least one of the two horizontal borders $\mathcal{S}(x_s, \varepsilon)$ or $\{y_{s+\delta} \colon\ y \in S(x_s, \varepsilon)\}$, of a rectangle $R(x_s, \varepsilon, \delta) \in \mathcal{R}$. Hence,

$$\#\{0 \leqslant j < N \colon\ \mathcal{S}_{N,j} \text{ is not properly covered by } \mathcal{R}\} \leqslant 2\#\mathcal{R}.$$

Projecting the statistical sum. According to Lemma 10.2, and taking into account the specification property, we have

$$\tau_\Phi\big(R(x_s,\varepsilon,\delta)\big) \leqslant \max_{z\in R_X(x_s,\varepsilon)} \sum_{j=0}^{\tau} \phi(\sigma^j z) + \bar{a} \\ \leqslant \phi\big(R_X(x_s,\varepsilon)\big) + \bar{a} + n_0 \max u.$$

Let us assume that $\mathcal{R}$ is such that

$$\max \operatorname{diam}\big(R(x_s,\varepsilon,\delta)\big) := \varepsilon_0 < 1/3.$$

Since $\varepsilon,\delta < \operatorname{diam}(R(x_s,\varepsilon,\delta))$ then $\varepsilon + 2\delta < 1$ for each $R(x_s,\varepsilon,\delta) \in \mathcal{R}$. Now, let $\mathcal{R}' := \{R(x_s,\varepsilon,\delta) \in \mathcal{R}\colon s < 1-\varepsilon_0\}$. With all the previous estimates we obtain,

$$\mathcal{M}(X^\phi,\alpha,q,\mathcal{R}) \geqslant \sum_{R(x_s,\varepsilon,\delta)\in\mathcal{R}'} \big\{\operatorname{diam}\big(R(x_s,\varepsilon,\delta)\big)\big\}^\alpha e^{-q\tau_\Phi(R_X(x_s,\varepsilon))} \\ \geqslant C_0 \sum_{j\in P_N} \sum_{R(x_s,\varepsilon,\delta)\in\mathcal{R}_{N,j}} \frac{\delta \times e^{-(\alpha+1)u(R_X(x_s,\varepsilon))-q\phi(R_X(x_s,\varepsilon))}}{\#\{0\leqslant j<N\colon R(x_s,\varepsilon,\delta)\in\mathcal{R}_{N,j}\}} \\ \geqslant \frac{C_0}{N} \sum_{j\in P_N} \sum_{R(x_s,\varepsilon,\delta)\in\mathcal{R}_{N,j}} e^{-(\alpha+1)u(R_X(x_s,\varepsilon))-q\phi(R_X(x_s,\varepsilon))},$$

where $C_0 := \exp(-q(n_0 \max u + \bar{a}))$ and

$$P_N := \big\{0 \leqslant j < \lfloor N(1-\varepsilon_0)\rfloor\colon \mathcal{S}_{N_j} \text{ is properly covered by} \mathcal{R}\big\}.$$

As was mentioned before, there are at most $2\#\mathcal{R}$ indices j, $0\leqslant j<N$, such that $\mathcal{S}_{N,j}$ is not properly covered by $\mathcal{R}$, hence $\#P_N \geqslant N(1-\varepsilon_0) - 2\#\mathcal{R} - 1$. From this we obtain

$$\mathcal{M}(X^\phi,\alpha,q,\mathcal{R}) \\ \geqslant C_0\bigg(1-\varepsilon_0-\frac{2\#\mathcal{R}+1}{N}\bigg) \min_{j\in P_N} \sum_{c\in\mathcal{C}_{N,j}} e^{-(\alpha+1)u(c)-q\phi(c)}.$$

Let $\mathcal{C}_\mathcal{R} := \{R_X(x_s,\varepsilon)\colon R(x_s,\varepsilon,\delta)\in\mathcal{R}\}$ the collection containing all the horizontal projections of rectangles in $\mathcal{R}$. Notice that $\mathcal{C}_\mathcal{R}$ is a cover of X by cylinder sets, and that $\mathcal{C}_{N,j} \subset \mathcal{C}_\mathcal{R}$ is a subcover for each N and $j\in P_N$. With this we have,

$$\mathcal{M}(X^\phi,\alpha,q,\mathcal{R}) \geqslant C_0\bigg(1-\varepsilon_0-\frac{2\#\mathcal{R}+1}{N}\bigg) \min_{\mathcal{C}\subset\mathcal{C}_\mathcal{R}} \sum_{c\in\mathcal{C}} e^{-(\alpha+1)u(c)-q\phi(c)},$$

where the minimum is taken over all subcovers of X by cylinder sets. This holds for arbitrary N, hence, taking $N \geqslant 2\#\mathcal{R}/\varepsilon_0$ we obtain

$$\mathcal{M}\big(X^\phi, \alpha, q, \mathcal{R}\big) \geqslant C_0(1-2\varepsilon_0) \min_{\mathcal{C}\subset\mathcal{C}_\mathcal{R}} \sum_{c\in\mathcal{C}} e^{-(\alpha+1)u(c)-q\phi(c)}.$$

Relating it to the topological pressure. Proposition 10.1 ensures that the X-projection of the rectangle $R(x_s, \varepsilon, \delta)$ is the cylinder set $\zeta^{m(x_s,\varepsilon)}(x)$, with $m(x_s, \varepsilon) \to \infty$ as $\varepsilon \to 0$. Indeed, it is easy to derive from the computations we did in the proof of Proposition 10.1 that if $\varepsilon \leqslant \varepsilon_0$, then, irrespectively of the value of s,

$$m(x_s, \varepsilon) \geqslant m(\varepsilon_0) := \left\lceil \frac{\log(\exp(\min u) - 1) - \log(\varepsilon_0)}{\min u} \right\rceil.$$

Therefore, as we consider covers of X^ϕ by rectangles whose diameter decreases to zero, we obtain

$$\mathcal{M}\big(X^\phi, \alpha, q\big) \geqslant C_0 \times \lim_{m\to\infty} \inf\big\{\mathcal{Z}\big(0, \big((1-\alpha)u - q\phi\big), \mathcal{C}, X\big) \colon |\mathcal{C}| \geqslant n\big\},$$

with $\mathcal{Z}$ as defined in (2.16). According to what was exposed therein, the limit in the right hand side of the previous inequality diverges as long as $P_X((1-\alpha)u - q\phi|\sigma) > 0$, i.e., $P_X((1-\alpha)u - q\phi|\sigma) > 0$ implies that $\mathcal{M}(X^\phi, \alpha, q) = \infty$, and with this we have proved the claim. □

As was mentioned above, Theorem 10.1 follows directly from Claims 10.1 and 10.2.

PART IV

MEASURE THEORETICAL RESULTS

Chapter 11

Invariant Measures and Poincaré Recurrences

In this chapter we will make use of an ergodic invariant measure μ (one can find main results about such measures in [26,45,49,104,106] or other books). We consider a pointwise dimension for Poincaré recurrences, similar to the pointwise dimension for a measure introduced by L.-S. Young [124], and show that it exists almost every where and is equal to a number independent of the point. Then we show that this number is, in fact, the dimension for a measure. On the way to prove this result we study local rates for Poincaré recurrences which were introduced independently in [5] and [113]. Chapter 13 is devoted to a proof of the variational principle which is possible to fulfill mainly because of the validity of the Bowen-type equation for the spectrum of dimensions for Poincaré recurrences.

11.1. Pointwise dimension and local rates

The framework will be the same as in Chapter 10, i.e., (Ω, σ) is a specified subshift of Ω_p, endowed with the ultrametric

$$d_\Omega(\omega, \varpi) := e^{-u(\zeta^n(\omega))}, \quad \forall \varpi \in \zeta^n(\omega) \setminus \zeta^{n+1}(\omega) \tag{11.1}$$

which as we already mentioned, is compatible with the product topology on Ω_p. Let us just remind that $u : \Omega \to \mathbb{R}^+$ is a Hölder continuous function, and that

$$u\big(\zeta^n(\omega)\big) := \max_{\varpi \in \zeta^n(\omega)} \sum_{j=0}^{n-1} u\big(\sigma^j \varpi\big). \tag{11.2}$$

As in Chapters 5 and 6, the specified subshift will be used as the model for a geometric construction as those introduced in Chapter 3. Additionally, we consider in Ω a Borel probability measure μ, which we suppose to be ergodic with respect to the shift. The q-pointwise dimension (for Poincaré recurrences) of the measure μ at the point $\omega \in \Omega$ is defined by

$$d_{\mu,q}(\omega) := \lim_{\varepsilon \to 0^+} \inf_{\varpi \in B(\omega,\varepsilon)} \frac{\log \mu(B(\varpi, \varepsilon)) + q\tau(B(\varpi, \varepsilon))}{\log \varepsilon}, \tag{11.3}$$

whenever this limit exists. Below we will show that this limit exists and it is constant in a set of full measure μ, when μ has positive entropy with respect to the shift. We already proved in Section 3.1.1 that the open ball $B(\omega, \varepsilon)$ is nothing else but the cylinder set $\zeta^{n_{\omega,\varepsilon}}(\omega)$, where

$$n_{\omega,\varepsilon} := \min\{n \in \mathbb{N}: e^{-u(\zeta^n(\omega))} < \varepsilon\}.$$

Furthermore, since d_Ω is a ultrametric, $B(\varpi, \varepsilon) = B(\omega, \varepsilon)$ for each $\varepsilon > 0$ and every $\varpi \in B(\omega, \varepsilon)$. Therefore

$$d_{\mu,q}(\omega) := \lim_{\varepsilon \to 0+} \frac{\log \mu(\zeta^{n_{\omega,\varepsilon}}(\omega)) + q\tau(\zeta^{n_{\omega,\varepsilon}}(\omega))}{\log \varepsilon}, \tag{11.4}$$

whenever the limit exists.

The proof of the existence of a local spectrum of dimensions for Poincaré relies essentially on two facts:

(a) the Shannon–McMillan–Breiman theorem, and
(b) the existence of local rates of return times.

The first fact will be discussed in details in Chapter 19 where a proof, not based on the martingale convergence theorem, will be provided. (See also [91].) Due to its novelty and its relevance for the subject matter of this book we devote more attention to the fact (b), and to some of its consequences. We will do this in Sections 11.4–11.5.

Given $q \in \mathbb{R}$, the q-dimension of Poincaré recurrences of the measure μ is give by

$$\alpha^{\mu}(q) := \inf\{\alpha(\Omega', q): \Omega' \subset \Omega \text{ and } \mu(\Omega') = 1\}. \tag{11.5}$$

In this way, the spectrum of dimensions for Poincaré recurrences of the measure μ is the function $q \mapsto \alpha^{\mu}(q)$. This definition applies directly to the general framework of discrete dynamical systems supplied with a Borel probability measure.

Following the approach of Pesin [97] and Chazottes and Saussol [39], we will prove in Section 12 that $\alpha^{\mu}(q)$ coincides with the constant value of $d_{\mu,q}(\omega)$ in a set of full measure μ.

The nodal point of this chapter is that, at least in the framework we consider here, the local dimension $d_{\mu,q}(\omega)$ may be computed. Indeed, thanks to Theorem 11.3, the Birkhoff's individual ergodic theorem (stated and proved in Chapter 18), and the Shannon–McMillan–Breiman theorem (Chapter 19), we can compute $d_{\mu,q}(\omega)$ and relate it to the metric entropy and to the Lyapunov exponent of the measure. We will do it in Section 11.6.

11.2. The Shannon–McMillan–Breiman theorem

Here we are considering μ to be an ergodic probability measure on Ω. The entropy of μ with respect to the shift is the limit

$$h(\mu) := -\lim_{n\to\infty} \frac{1}{n} \sum_{\zeta^n(\omega)\in\zeta^n} \mu\big(\zeta^n(\omega)\big) \log \mu\big(\zeta^n(\omega)\big), \tag{11.6}$$

whose existence is ensured by subadditivity (see [120] for instance), and it is always finite for a subshift $\Omega \subset \Omega_p$. The Shannon–McMillan–Breiman theorem establishes that

$$\mu\left\{\omega \in \Omega\colon \lim_{n\to\infty} \frac{-\log \mu(\zeta^n(\omega))}{n} = h(\mu)\right\} = 1. \tag{11.7}$$

As usual when for a property P we have $\mu\{\omega \in \Omega\colon P(\omega)\} = 1$, we will alternatively use the phrases "$P(\omega)$ holds μ-almost everywhere", or "$P(\omega)$ for μ-almost all $\omega \in \Omega$", or "$P(\omega)$ for all ω in a set of full measure μ".

11.3. Kolmogorov complexity and Brudno's theorem

Let M be a Turing machine on the alphabet of p symbols. Let $\underline{\omega}$ be a finite word, prefix of $\omega \in \Omega_p$ of the length $n = |\underline{\omega}|$. The Kolmogorov complexity $K_M(\underline{\omega})$ of $\underline{\omega}$ with respect to M is the length of the shortest program such that it yields the output $\underline{\omega}$ for the input $n \in \mathbb{N}$. For any universal Turing machine U there exists a constant $C > 0$ such that

$$K_U(\underline{\omega}) \leqslant K_M(\underline{\omega}) + C.$$

For $\omega \in \Omega_p$, the average upper complexity is the upper limit

$$\overline{K}(\omega) := \limsup_{|\underline{\omega}|\to\infty} \frac{K_U(\underline{\omega})}{|\underline{\omega}|}. \tag{11.8}$$

Note that by the previous inequality the upper average complexity does not depend on the choice of the universal Turing machine. The lower average complexity $\underline{K}(x)$ is defined in a analogous way, replacing in (11.8) lim sup by lim inf. For details, see for instance [79].

In [35] Brudno establishes the equality between the upper average complexity and the metric entropy, which in our context may be paraphrased as follows.

THEOREM 11.1. *For ω in a set of full measure μ we have*

$$\overline{K}(\omega) = h(\mu).$$

In his Ph.D. work White extended Brudno's results. An adapted version of one of White's results, which immediately follows from Proposition 2.5 in [121], is the following.

THEOREM 11.2. *For ω in a set of full measure μ we have*

$$\underline{K}(\omega) = \overline{K}(\omega) = h(\mu).$$

We will use this theorem in the second proof of Theorem 11.3 below.

11.4. The local rate of return times

For a specified subshift (Ω, σ) such that $h(\sigma | \Omega) > 0$, one has

$$\lim_{n\to\infty} \frac{\tau(\zeta^n(\omega))}{n} = 1. \tag{11.9}$$

Below we will expose two alternative proofs of this result. The first one is based on the SMB theorem and inspired on the proof presented in [7]. Our second proof is taken, almost unchanged, from [113]. It is based on Brudno's theorem relating the Kolmogorov complexity to the metric entropy.

THEOREM 11.3. *Let (Ω, σ) be a specified subshift of the full shift on p symbols, and let μ be a Borel probability measure on (Ω, d_Ω). Suppose that μ is σ-ergodic and such that $h(\mu) > 0$, then*

$$\lim_{n\to\infty} \frac{\tau(\zeta^n(\omega))}{n} = 1$$

for μ-almost every $\omega \in \Omega$.

As an example of calculating local rates see [77].

11.4.1. *Proof of Theorem 11.3 based on the SMB Theorem*

Since (Ω, σ) is specified, it immediately follows that $\tau(\zeta^n(\omega)) \leqslant n + n_0$, where n_0 is the specification length. Then $\limsup_n \tau(\zeta^n(\omega))/n \leqslant 1$ for all $\omega \in \Omega$, which in particular implies that

$$\mu\left\{\omega \in \Omega\colon \limsup_{n\to\infty} \frac{\tau(\zeta^n(\omega))}{n} \leqslant 1\right\} = 1.$$

With this we conclude the *limsup-part* of the proof.

The Egorov theorem ensures that for each $\varepsilon > 0$ there is a Borel set $X_\varepsilon \subset \Omega$ with measure $\mu(X_\varepsilon) > 1-\varepsilon$, such that $-\log \mu(\zeta^n(\omega))/n$ converges to $h := h(\mu)$ uniformly on X_ε.

Fix $\varepsilon > 0$, $0 < \delta < h$ and $C = C(\varepsilon, \delta) \geqslant 1$ such that for each $\omega \in X_\varepsilon$ and each $n \in \mathbb{N}$ we have

$$C^{-1} e^{-(h+\delta)n} \leqslant \mu\big(\zeta^n(\omega)\big) \leqslant C e^{-(h-\delta)n}. \tag{11.10}$$

The constant C is needed to ensure that these inequalities hold for small n.

The strategy of the *liminf-part* of the proof is the following. For each $n \in \mathbb{N}$ we fix the set

$$X^\delta_{\varepsilon,n} := \left\{\omega \in X_\varepsilon \colon \tau\big(\zeta^n(\omega)\big) \leqslant \frac{h-\delta}{h+\delta}\left(n - \frac{2\log(n)}{h-\delta}\right)\right\}$$

and then prove that the series $\sum_{n=1}^\infty \mu(X^\delta_{\varepsilon,n})$ converges. According to the Borel–Cantelli lemma (see for instance [26]) this result implies that $\mu(\limsup_n X^\delta_{\varepsilon,n}) = 0$, or equivalently

$$\mu\big(\liminf_n\big(X_\varepsilon \setminus X^\delta_{\varepsilon,n}\big)\big) = \mu(X_\varepsilon) \geqslant 1 - \varepsilon.$$

This inequality implies then that for all $\varepsilon > 0$ and $0 < \delta < h$

$$\mu\left(\omega \in \Omega \colon \liminf_{n\to\infty} \frac{\tau(\zeta^n(x))}{n} \geqslant \frac{h-\delta}{h+\delta}\right) \geqslant 1 - \varepsilon.$$

Thus, to conclude the proof all we are left to do is to show that $\sum_{n=1}^\infty \mu(X^\delta_{\varepsilon,n})$ converges. For each n and $k \in \mathbb{N}$ consider the collection of cylinder sets

$$R_{n,k} := \big\{c \in \zeta^n \colon c \cap X_\varepsilon \neq \emptyset \text{ and } \tau(c) = k\big\}.$$

Using the inequalities (11.10) derived from the SMB theorem we obtain

$$\begin{aligned}
\mu\big(X^\delta_{\varepsilon,n}\big) &\leqslant \sum_{k=1}^{\lfloor \frac{h-\delta}{h+\delta}(n - \frac{2\log(n)}{h-\delta})\rfloor} \sum_{c\in R^k_n} \mu(c) \\
&\leqslant C \sum_{k=1}^{\lfloor \frac{h-\delta}{h+\delta}(n - \frac{2\log(n)}{h-\delta})\rfloor} \sum_{c\in R^k_n} e^{-n(h-\delta)} \\
&\leqslant C^2 \sum_{k=1}^{\lfloor \frac{h-\delta}{h+\delta}(n - \frac{2\log(n)}{h-\delta})\rfloor} e^{-n(h-\delta)+k(h+\delta)} \left(C^{-1} \sum_{c\in R^k_n} e^{-k(h+\delta)}\right).
\end{aligned}$$

Since $\tau(c) = k$ for each $c \in R^k_n$, then every $\omega \in c \in R^k_n$ satisfies $\omega_i = \omega_{i+k}$ for $0 \leqslant i < n - k - 1$. Thus, a cylinder $c = [\underline{\omega}] \in R^k_n$ is completely determined by the prefix $\underline{\omega}$ of length k of any $\omega \in c$. Furthermore, if $c \cap X_\varepsilon \neq \emptyset$, then

$\zeta^k(\omega) \cap X_\varepsilon \neq \emptyset$ for any $\omega \in c$. Because of this, the last inequality implies

$$\mu\left(X^\delta_{\varepsilon,n}\right) \leqslant C^2 \sum_{k=1}^{\lfloor \frac{h-\delta}{h+\delta}(n-\frac{2\log(n)}{h-\delta})\rfloor} e^{-(n-k\frac{h+\delta}{h-\delta})(h-\delta)} \left(C^{-1} \sum_{c\in\zeta^{k,\varepsilon}} e^{-k(h+\delta)}\right),$$

where $\zeta^{k,\varepsilon} := \{c \in \zeta^k\colon\ c \cap X_\varepsilon \neq \emptyset\}$. Using once again (11.10) we obtain

$$\begin{aligned}\mu\left(X^\delta_{\varepsilon,n}\right) &\leqslant C^2 \sum_{k=1}^{\lfloor \frac{h-\delta}{h+\delta}(n-\frac{2\log(n)}{h-\delta})\rfloor} e^{-(n-k\frac{h+\delta}{h-\delta})(h-\delta)} \\ &\leqslant C^2 \sum_{\ell=\lceil \frac{2\log(n)}{h-\delta}\rceil}^{\infty} e^{-\ell(h-\delta)} \leqslant \left(\frac{e^{h-\delta}C^2}{1-e^{-(h+\delta)}}\right)\frac{1}{n^2},\end{aligned}$$

from which it follows that $\sum_{n=1}^{\infty} \mu(X^\delta_{\varepsilon,n}) < \infty$. Thus, as mentioned above, for each $\varepsilon > 0$ and $0 < \delta < h$ we have

$$\mu\left(\omega \in X\colon\ \liminf_{n\to\infty} \frac{\tau(\zeta^n(\omega))}{n} \geqslant \frac{h-\delta}{h+\delta}\right) \geqslant 1-\varepsilon.$$

The result follows from the fact ε and δ may be taken arbitrarily small.

11.4.2. Proof of Theorem 11.3 based on Brudno's Theorem

The *limsup-part* of the proof is exactly the same as in the previous proof. For the *liminf-part* we have the following. Fix the universal Turing machine U, reading input sequences $\omega \in \Omega$ on p symbols. For any point $\omega \in \Omega$ such that $\tau(\zeta^n(\omega)) = k\}$ we have that $\omega_i = \omega_{i+k}$ for each $0 \leqslant k < n-k$. Therefore in order to output the prefix $\underline{\omega}(n)$ of length n it is enough to compute the prefix $\underline{\omega}(k)$ of length k and then output this prefix repeatedly until a word of length n is completed. Let us denote by $[n]$ the finite word of p-digits encoding the natural number n in the base p. Then, there is a constant $C > 0$ such that

$$K_U\left(\underline{\omega}(n)\right) \leqslant K_U\left(\underline{\omega}(k)\right) + K_U([n]) + C \quad \forall\, x \in R^k_n.$$

Here C stands for the length of the section of the program which orders the repetition of the prefix of length k and $R^k_n := \{\omega \in \Omega\colon\ \tau(\zeta^n(\omega)) = k\}$. On the other hand $K_U([n]) \leqslant C' + \log_p(n)$, which corresponds to the worst case of a program (encoded in a sequence of length C') outputing directly the word $[n]$, whose length is bounded by $\log_p(n)$. Therefore, there is a positive constant $C'' > 0$ such that

$$K_U\left(\underline{\omega}(n)\right) \leqslant K_U\left(\underline{\omega}(k)\right) + \log_p(n) + C'' \quad \forall\, \omega \in R^k_n.$$

From this it follows

$$\begin{aligned}\underline{K}(\omega) &\leqslant \liminf_{n\to\infty} \frac{K_U(\underline{\omega}(\tau(\zeta^n(\omega)) - 1)) + \log_p(n)}{n} \\ &= \liminf_{n\to\infty} \frac{K_U(\underline{\omega}(\tau(\zeta^n(\omega)) - 1))}{\tau(\zeta^n(\omega))} \frac{\tau(\zeta^n(\omega))}{n} \\ &\leqslant \limsup_{n\to\infty} \frac{K_U(\underline{\omega}(n-1))}{n} \liminf_{n\to\infty} \frac{\tau(\zeta^n(\omega))}{n} \\ &= \overline{K}(\omega) \liminf_{n\to\infty} \frac{\tau(\zeta^n(\omega))}{n}.\end{aligned}$$

By using Theorem 11.2 we obtain

$$\overline{K}(\omega) \leqslant \overline{K}(\omega) \liminf_{n\to\infty} \frac{\tau(\zeta^n(\omega))}{n},$$

for all ω in a set of full measure μ, which implies the *liminf-part*, namely

$$\frac{\tau(\zeta^n(\omega))}{n} \geqslant 1,$$

for all ω in a set of full measure μ.

We studied local rates for specified systems with positive entropy. The situation is different for non-chaotic cases. We consider here one of simplest such systems.

11.4.3. Rotations of the circle

Consider a rotation $f_\nu : x \mapsto x - \nu$, mod 1 (i.e., $f_\nu^{-1}x = x + \nu$, mod 1), on the circle $S^1 = \{x, \text{mod} 1\}$, where $0 < \nu < 1$ is an irrational number. The number ν can be approximated by rational numbers p/q (p and q are relatively prime) in such a way that

$$\left|\nu - \frac{p}{q}\right| < \frac{1}{q^{\beta+1}} \tag{11.11}$$

for some value β and some pair (p, q). Let $\beta(\nu) := \sup \beta$ where the supremum is taken over all β for which inequality (11.11) has infinitely many solutions (p, q) with $q > 0$. Assume that $\beta(\nu) < \infty$, i.e., ν is a Diophantine number. Then for every $\delta \in (0, 1)$ the inequality

$$\left|\nu - \frac{p}{q}\right| < \frac{1}{q^{\beta(\nu)+1-\delta}} \tag{11.12}$$

holds for infinitely many relatively prime pairs (p_i, q_i), with $q_i \to \infty$, as $i \to \infty$. Consider a partition ζ^0 of S^1 made with two closed intervals $[0, \omega]$ and $[\omega, 1]$, and let $\zeta^n := \bigvee_{i=0}^{n-1} f_\nu^{-i}\zeta^0$. Denote by $\zeta^n(x)$ an element of ζ^n containing a point x (the definition is correct for all x except for a point belonging to the set of the endpoints of intervals in ζ^n). The rotation is metrically isomorphic to the subshift $\mathrm{clos}(\pi([0,1)))$, where the coding map $\pi : [0,1] \to \{0,1\}^{\mathbb{N}}$ is defined in the following way: $\pi(x)_n = 0$ is $f_\nu^n(x) \in [0,\nu)$ and $\pi(x)_n = 1$ if $f_\nu^n(x) \in [\nu, 1)$. We now state the following.

THEOREM 11.4. ([5]) *If* $\beta(\nu) > 3$ *then*

$$\liminf_{n\to\infty} \frac{1}{n}\tau\bigl(\zeta^n(x)\bigr) = 0 \tag{11.13}$$

for almost every x *with respect to the Lebesgue measure on* S^1.

PROOF. Start by choosing δ to be so small that

$$\beta(\nu) - \delta > 3. \tag{11.14}$$

Without loss of generality we may assume that

$$\left|\nu - \frac{p_i}{q_i}\right| = \inf_{p\in\mathbb{Z}}\left|\nu - \frac{p}{q_i}\right|.$$

Then because of (11.12), we have

$$\mathrm{dist}\bigl(x, f_\nu^{q_i}x\bigr) = \inf_{p\in\mathbb{Z}} |x + q_i\nu - x - p| = q_i\left|\nu - \frac{p_i}{q_i}\right| < \frac{1}{q_i^{\beta(\nu)-\delta}}. \tag{11.15}$$

Now introduce a number $\alpha > 1$ such that $1 + 2\alpha < \beta(\nu) - \delta$. Let $m_i := [q_i^{1+\alpha}]$ the integer part of $q_i^{1+\alpha}$ and let

$$A_{m_i} := \#\left\{\zeta^{m_i}(x)\colon \mathrm{diam}\,\zeta^{m_i}(x) \geqslant \frac{1}{q_i^{1+2\alpha}}\right\},$$

$$B_{m_i} := \#\left\{\zeta^{m_i}(x)\colon \mathrm{diam}\,\zeta^{m_i}(x) < \frac{1}{q_i^{1+2\alpha}}\right\}.$$

Since $A_{m_i} + B_{m_i} = 2m_i$ then $A_{m_i} + B_{m_i} \leqslant 2q_i^{1+\alpha}$. This inequality implies that $B_{m_i} \leqslant 2q_i^{1+\alpha}$. Moreover,

$$\mu(\mathcal{B}_{m_i}) \leqslant 2q_i^{1+\alpha}\cdot q_i^{-(1+2\alpha)} = 2q_i^{-\alpha},$$

where

$$\mathcal{B}_{m_i} := \bigcup_{\mathrm{diam}\,\zeta^{m_i}(x) < q_i^{-(1+2\alpha)}} \zeta^{m_i}(x),$$

the union of the elements of the partition ζ^{m_i} of small diameter. In view of Borel–Cantelli Lemma (recall that $\alpha > 1$) – see for instance [26], we have that μ-almost every point x belongs to the complement of $\mathcal{B}_{m_i}$ provided that m_i is large enough, i.e., $\operatorname{diam} \zeta^{m_i}(x) \geqslant q_i^{-(1+2\alpha)}$.

Now because of (11.15) and the assumption $\beta(\nu) > 3$, we have

$$\tau(\zeta^{m_i}(x)) \leqslant q_i.$$

Therefore,

$$\liminf \frac{\tau_{f_\nu}(\zeta^n(x))}{n} \leqslant \lim_{i\to\infty} \frac{q_i}{q_i^{1+\alpha}} = 0$$

Lebesgue-almost everywhere. □

11.5. Remarks on local rates

REMARK 11.1. Following exactly the same lines as for Theorem 11.3 above, one can prove the following.

THEOREM 11.5. *Given a discrete dynamical system (X, T), a finite Borel partition ζ, and an ergodic Borel probability measure μ. Let the metric entropy of T with respect to ζ (defined in exact analogy to* (11.6)) *be positive. Then*

$$\liminf_{n\to\infty} \frac{\tau(\zeta^n(x))}{n} \geqslant 1$$

for μ almost all x.

In Section 11.4.3 we considered rotations $x \mapsto x+\alpha \pmod 1$ in the unite circle S^1. We proved therein that, for $\zeta := \{[0, 1-\alpha), [1-\alpha, 1)\}$, it could happen that

$$\liminf_{n\to\infty} \frac{\tau(\zeta^n(x))}{n} = 0$$

for all x in a set of full Lebesgue measure. The positivity of the entropy is therefore a unavoidable condition for the lower local rate to be greater or equal to 1. This fact suggests that the converse of Theorem 11.5 may hold, however this was disproved in [38]. Instead it was proved the following partial converse.

THEOREM 11.6. *Given a discrete dynamical system (X, T), a finite Borel partition ζ, and an ergodic Borel probability measure μ, let*

$$\liminf_{n\to\infty} \frac{\tau(\zeta^n(x))}{n} \geqslant 1$$

for μ almost all x and every non-trivial (made of at least two atoms, all of them *with positive measure) measurable partition ζ. Then μ is weakly mixing with respect to T, i.e.,*

$$\lim_{n\to\infty} \frac{1}{n} \sum_{i=0}^{n-1} \left|\mu\left(A \cap T^{-i} B\right) - \mu(A)\mu(B)\right| = 0,$$

for every pair A, B of Borel sets.

Relationships between mixing and the positiveness of entropy have been widely studied, and the general situation here is that some property concerning the entropy implies some kind of mixing. For instance, *for a discrete dynamical system (X, T) and μ an ergodic Borel probability measure, let the metric entropy of T with respect to every non-trivial partition ζ be positive (i.e., (X, T, μ) is a K-system), then (X, T) is strongly mixing with respect to μ, i.e., $\lim_{n\to\infty} \mu(A \cap T^{-n} B) = \mu(A)\mu(B)$ for every pair A, B of Borel sets.* See for instance [120].

REMARK 11.2. In some special cases, by using the lower bound given by Theorem 11.5, it is possible to derive a relationship between the local rate of return times and the Lyapunov exponent of an invariant measure. In [113] such a relationship is established for a special class of piecewise monotonic transformations including expanding Markov maps. In the particular case of a piecewise expanding Markov map T (which we already studied in Chapter 9) and μ an ergodic Borel probability measure with positive entropy, the equality

$$\lim_{r\to 0} \frac{\tau(B(x,r))}{\log(1/r)} = \frac{1}{\lambda_\mu}$$

holds for μ almost all $x \in X$. Here $\lambda_\mu := \int |T'(x)|\, d\mu(x)$ is the Lyapunov exponent of the measure μ. In [114], the same authors generalize such kind of relationship between local rates and Lyapunov exponents to higher dimensional transformations.

REMARK 11.3. The local rate of return times has also appeared in the study of statistics of return times. In the same framework, and keeping the same notations as in the previous remarks, let

$$F_{\zeta^n(x)}(t) := \frac{\mu\{x' \in \zeta^n(x)\colon \tau(x', \zeta^n(x)) > t/\mu(\zeta^n(x))\}}{\mu(\zeta^n(x))}. \tag{11.16}$$

The Kac's theorem justifies the chosen normalization. Indeed, this theorem states that $\int_{\zeta^n(x)} \tau(x', \zeta^n(x))\, d\mu(x') = 1$, so that the mean return time to $\zeta^n(x)$ diverges like $\mu(\zeta^n(x))^{-1}$ as n goes to infinity. The asymptotic behavior of the law $F_{\zeta^n(x)}$ has been studied since the early 1990s [103,67], and it was found to be exponential for mixing systems. To be more precise, one can be derive from Theorem 2.1 and Lemma 2.4 in [68] the following.

THEOREM 11.7. *Let (X, T) be a discrete dynamical system, ζ a finite or countable Borel partition, and μ a Borel probability measure. If the partition is α-mixing, then*

$$\lim_{n\to\infty} \sup_{t>0} \left| F_{\zeta^n(x)}(t) - e^{-t} \right| = 0$$

for any non-trivial finite or countable partition ζ, and for μ-almost all $x \in X$.

Let us remind that the partition ζ is said to be α-mixing with respect to μ if the function

$$\alpha(n) := \sup_{k,\ell} \sup_{c\in\zeta^k,\ c'\in\zeta^\ell} \left| \frac{\mu(c \cap T^{-(n+k)}c')}{\mu(c)} - \mu(c') \right| \to 0 \quad \text{as } n \to 0. \tag{11.17}$$

The supremum is taken over cylinders c such that $\mu(c) > 0$.

The convergence rate in (11.7) can be related to the local rate of return times in the following way. For $x \in X$ let

$$\underline{\beta}(x) := \liminf_{n\to\infty} \frac{\log(\sup_{t>0} |F_{\zeta^n(x)}(t) - e^{-t}|)}{\log \mu(\zeta^n(x))}, \tag{11.18}$$

and define $\overline{\beta}(x)$ in the analogous way, by replacing lim inf by lim sup. From Theorem 2.1 and Lemma 2.4 in [68], and following the idea of the proof of Theorem 3 in [112], we can prove the following.

THEOREM 11.8. *Let (X, T) be a discrete dynamical system, ζ a finite or countable Borel partition, and μ an ergodic Borel probability measure. If the metric entropy of μ with respect to ζ is positive, ζ is α-mixing with α vanishing exponentially fast, and (X, T) satisfies the specification assumption, then*

$$\underline{\beta}(x) = \overline{\beta}(x) = \lim_{n\to\infty} \frac{\tau(\zeta^n(x))}{n}$$

for μ-almost every $x \in X$.

11.6. The q-pointwise dimension for Poincaré recurrences

Let us come back to the basic framework of this part of the book. Remind that we are dealing with a specified subshift (Ω, σ), supplied with an ultrametric d_Ω defined from a Hölder continuous function $u : \Omega \to \mathbb{R}^+$ (see Eq. (11.1)). We are also dealing with an ergodic Borel probability measure μ such that $h(\mu)$, the metric entropy we defined in (11.6), is positive.

Associated to the ultrametric d_Ω we define the Lyapunov exponent of the measure μ,

$$\lambda_\mu := \int_\Omega u(\omega)\, d\mu(\omega). \tag{11.19}$$

This definition is in agreement with the traditional notion of Lyapunov exponent. To see this, think of (Ω, σ) as a model for a piecewise differentiable map $T : [0, 1] \to [0, 1]$. Both systems would be related through a semiconjugacy $\chi : \Omega \to [0, 1]$ such that $|T'(\chi(\omega))| = \exp(u(\omega))$. Taking into account the Birkhoff's individual ergodic theorem we obtain the relations

$$\lim_{n\to\infty} \frac{1}{n} \log\left|T^{(n)\prime}\big(\chi(\omega)\big)\right| = \lim_{n\to\infty} \frac{1}{n} \sum_{j=0}^{n-1} u\big(\sigma^j\omega\big) = \int_\Omega u(\omega)\, d\mu(\omega).$$

Therefore, by using the same notation μ for both the original measure in Ω and the one induced by χ in $[0, 1]$, we may write

$$\lim_{n\to\infty} \frac{1}{n} \log\left|T^{(n)\prime}(x)\right| = \lambda_\mu$$

for μ-almost all $x \in [0, 1]$.

We have already all the tools needed to establish a formula for the q-pointwise dimension for Poincaré recurrences for the measure μ. Let us put this in the form of a theorem.

THEOREM 11.9. *Let (Ω, σ), d_Ω, and μ be as above. Then the limit*

$$d_{\mu,q}(\omega) := \lim_{\varepsilon\to 0} \frac{\log \mu(B(\omega, \varepsilon)) + q\tau(B(\omega, \varepsilon))}{\log \varepsilon}$$

exists and it is such that

$$d_{\mu,q}(\omega) = \frac{h(\mu) - q}{\lambda_\mu}$$

for all ω in a set of full measure μ.

PROOF. As we already mention at the beginning of this section, the open ball $B(\omega, \varepsilon)$ coincides with the cylinder $\zeta^{n_{\omega,\varepsilon}}(\omega)$ with $n_{\omega,\varepsilon} := \min\{n \in \mathbb{N}\colon e^{-u(\zeta^n(\omega))} < \varepsilon\}$. On the other hand, since d_Ω is a ultrametric, $B(\varpi, \varepsilon) = B(\omega, \varepsilon)$ for each $\varepsilon > 0$ and every $\varpi \in B(\omega, \varepsilon)$. Therefore

$$d_{\mu,q}(\omega) := \lim_{\varepsilon\to 0^+} \frac{\log \mu(\zeta^{n_{\omega,\varepsilon}}(\omega)) + q\tau(\zeta^{n_{\omega,\varepsilon}}(\omega))}{\log(\varepsilon)},$$

whenever the limit exists. Theorem 11.3, the Shannon–McMillan–Breiman theorem, and the Birkhoff's individual ergodic theorem ensures that for μ-almost all $\omega \in \Omega$

$$\lim_{n\to\infty} \frac{\tau(\zeta^n(\omega))}{n} = 1,$$

$$\lim_{n\to\infty} \frac{-\log \mu(\zeta^n(\omega))}{n} = h(\mu), \quad \text{and}$$

$$\lim_{n\to\infty} \frac{1}{n}\sum_{j=0}^{n-1} u(\sigma^j\omega) = \int_{\Omega} u(\omega)\,d\mu(\omega) = \lambda_\mu.$$

Now, since $n_{\omega,\varepsilon} := \min\{n \in \mathbb{N}\colon e^{-u(\zeta^n(\omega))} < \varepsilon\}$, then

$$-\sup_{\varpi\in\zeta^{n_{\omega,\varepsilon}}(\omega)} \sum_{j=0}^{n_{\omega,\varepsilon}-1} u(\sigma^j\varpi) < \log(\varepsilon) \leqslant -\sup_{\varpi\in\zeta^{n_{\omega,\varepsilon}-1}(\omega)} \sum_{j=0}^{n_{\omega,\varepsilon}-2} u(\sigma^j\varpi).$$

For $u : \Omega \to \mathbb{R}^+$ a Hölder continuous function, there exists a constant $C > 0$ such that

$$\sup_{\varpi\in\zeta^{n_{\omega,\varepsilon}}(\omega)} \sum_{j=0}^{n_{\omega,\varepsilon}-1} u(\sigma^j\varpi) \leqslant \sum_{j=0}^{n_{\omega,\varepsilon}-1} u(\sigma^j\omega) + C,$$

therefore

$$-\left(\sum_{j=0}^{n_{\omega,\varepsilon}-1} u(\sigma^j\omega) - C'\right) < \log(\varepsilon) \leqslant -\left(\sum_{j=0}^{n_{\omega,\varepsilon}-1} u(\sigma^j\omega) + C'\right),$$

with $C' = \max(C, \max_{\omega\in\Omega} |u(\omega)|)$. From this we get

$$\begin{aligned}\frac{-\log \mu(\zeta^{n_{\omega,\varepsilon}}(\omega)) - q\tau(\zeta^{n_{\omega,\varepsilon}}(\omega))}{\sum_{j=0}^{n_{x,\varepsilon}-1} u(\sigma^j\omega) + C'} &\leqslant \frac{-\log \mu(\zeta^{n_{\omega,\varepsilon}}(\omega)) - q\tau(\zeta^{n^{\omega,\varepsilon}}(\omega))}{\log(\varepsilon)} \\ &\leqslant \frac{-\log \mu(\zeta^{n^{\omega,\varepsilon}}(\omega)) - q\tau(\zeta^{n_{\omega,\varepsilon}}(\omega))}{\sum_{j=0}^{n_{\omega,\varepsilon}-1} u(\sigma^j\omega) - C'}.\end{aligned}$$

Finally, since $\varepsilon \to 0^+$ is equivalent to $n_{\omega,\varepsilon} \to \infty$, then we obtain

$$\begin{aligned}\lim_{\varepsilon\to 0^+} \frac{\log \mu(\zeta^{n_{\omega,\varepsilon}}(\omega)) + q\tau(\zeta^{n_{\omega,\varepsilon}}(\omega))}{\log(\varepsilon)} &= \lim_{n\to\infty} \frac{-\frac{\log \mu(\zeta^n(\omega))}{n} - q\frac{\tau(\zeta^n(\omega))}{n}}{\frac{\sum_{j=0}^{n-1} u(\sigma^j\omega)}{n}} \\ &= \frac{h(\mu) - q}{\lambda_\mu},\end{aligned}$$

for μ-almost all $\omega \in \Omega$, and the proof is done. □

Chapter 12

Dimensions for Measures and q-Pointwise Dimension

12.1. Preliminaries and motivation

Theorems relating local dimensions and dimensions of measures have appeared in the literature at least since early 1980s. As far as we know, it was in [124] that Young first introduced the notion of Hausdorff dimension of measure, and proved a formula relating it to the Lyapunov exponents. More precisely, she proved the following.

THEOREM 12.1. *Let $F : M \to M$ be a C^2 diffeomorphism of a compact surface M, and let μ be an ergodic Borel probability measure with Lyapunov exponents $\lambda_1 \geqslant \lambda_2$. If $\lambda_1 \times \lambda_2 \neq 0$, then*

$$\dim_H(\mu) = h(\mu)\left(\frac{1}{\lambda_1} - \frac{1}{\lambda_2}\right).$$

It is worthwhile to remark that the idea behind the proof of this formula (already implicit in [23]) is just the same one we continue to exploit. The right hand side of the equation corresponds to a local dimension, which is proved to exist and to be constant in a set of full measure. This local dimension, when it exists, has to coincide with the dimension of the measure.

As a matter of motivation consider the following example. Let $\Omega \subset \Omega_p$ be a subshift. To each set $X' \subset \Omega$, we associate the Carathéodory dimension

$$h_{\text{top}}(X') := \sup\left\{\beta\colon \lim_{n\to\infty}\left(\inf\left\{\Xi(\beta, \mathcal{C}, X')\colon |\mathcal{C}| \geqslant n\right\}\right) = \infty\right\},$$

where

$$\Xi(\beta, \mathcal{C}, X') := \sum_{c\in\mathcal{C}} \exp\left(-\beta|c|\right),$$

$\mathcal{C}$ is a finite or countable cover by cylinders, and $|\mathcal{C}| \geqslant n$ means that all the cylinders c in $\mathcal{C}$ have length $|c|$ not smaller than n. This definition correspond to the dimension-like definition of topological pressure (Section 2.4.1) for the null potential.

Consider now a Borel probability measure μ, and define

$$h_{\text{top}}(\mu) := \inf\{h_{\text{top}}(X'):\ \mu(X') = 1\}.$$

The following result is a consequence of the Shannon–McMillan–Breiman theorem, and we present it here to illustrate the ideas of the main theorem of this section.

PROPOSITION 12.1. *Let $\Omega \subset \Omega_p$ be a subshift and μ a Borel probability measure, ergodic with respect to the shift. Then*

$$h_{\text{top}}(\mu) := h_\mu(T|\Omega).$$

Let us remark that, because of the Shannon–McMillan–Breiman theorem, the local dimension

$$h(\mu,\omega) := \lim_{n\to\infty} \frac{\mu(\zeta^n(\omega))}{n}$$

exists and takes the value, say $h(\mu)$, for μ-almost all ω.

PROOF. Let $X'' \subset \Omega$ be such that $\mu(X'') = 1$. Let $h := h(\mu)$ and define

$$\mathcal{G}_h := \{\omega \in \Omega:\ h(\mu,\omega) = h\}.$$

Shannon–McMillan–Breiman theorem ensure that $X' := X'' \cap \mathcal{G}_h$ has full measure.

The Egorov theorem together with the Shannon–McMillan–Breiman theorem implies that for each $\varepsilon > 0$ there exists $X'_\varepsilon \subset \Omega$ such that for each $\delta > 0$ there exists $n(\delta) \in \mathbb{N}$ such that

$$e^{-(h+\delta)n} \leqslant \mu(\zeta^n(\omega)) \leqslant e^{-(h-\delta)n}, \tag{12.1}$$

for all $n \geqslant n(\delta)$ and $\omega \in X'_\varepsilon$.

Fix $\delta > 0$ and let $\beta \leqslant h - \delta$. Then, for each cover $\mathcal{C}$ of X'_ε by cylinders of length $n \geqslant n(\delta)$ we have

$$\mathcal{Z}(\beta,\mathcal{C},X'_\varepsilon) = \sum_{c\in\mathcal{C}} e^{-\beta|c|} \geqslant \sum_{c\in\mathcal{C}} e^{-(h-\delta)|c|},$$

then, taking into account (12.1), we obtain that

$$\mathcal{Z}(\beta,\mathcal{C},X'_\varepsilon) \geqslant \sum_{c\in\mathcal{C}} \mu(c) \geqslant \mu(X'_\varepsilon) \geqslant 1-\varepsilon.$$

Since this is true for any cover $\mathcal{C}$ of X'_ε by cylinders of length $n \geqslant n(\delta)$, for all $\delta > 0$, then

$$\lim_{n\to\infty}\left(\inf\{\mathcal{Z}(\beta,\mathcal{C},X'_\varepsilon):\ |\mathcal{C}| \geqslant n\}\right) \geqslant 1-\varepsilon,$$

for $\beta < h$. From this it follows that $h_{\rm top}(X'_\varepsilon) \geqslant h$ for all $\varepsilon > 0$. Since $X' \supset X'_\varepsilon$, then $h_{\rm top}(X') \geqslant h$ as well.

For the converse inequality fix $\varepsilon > 0$ and $\delta > 0$ as before, and let $\beta \geqslant h + \delta$. Now, for each $n \geqslant n(\delta)$ let $\mathcal{C}_n$ be the cover of X'_ε by cylinders of length n. Since this cover is at the same time a partition, then by using (12.1) we obtain

$$\mathcal{Z}(\beta, \mathcal{C}_n, X'_\varepsilon) = \sum_{c\in\mathcal{C}_n} e^{-\beta n} \leqslant \sum_{c\in\mathcal{C}_n} e^{-(h+\delta)n} \leqslant \sum_{c\in\mathcal{C}_n} \mu(c) \leqslant \mu(X'_\varepsilon) \leqslant 1.$$

Therefore

$$\lim_{n\to\infty}\left(\inf\left\{\mathcal{Z}(\beta, \mathcal{C}, X'_\varepsilon)\colon\ |\mathcal{C}| \geqslant n\right\}\right) \leqslant \limsup_{n\to\infty} \mathcal{Z}(\beta, \mathcal{C}_n, X'_\varepsilon) \leqslant 1,$$

for all $\beta > h$. Therefore $h_{\rm top}(X'_\varepsilon) \geqslant h$ for all $\varepsilon > 0$. Let $\{\varepsilon_n\}_{n=1}^\infty$ be any positive sequence converging to zero, then $X' = \bigcup_{n=1}^\infty X'_{\varepsilon_n}$. By a general property of Carathéodory dimensions (see [97]) we have

$$h_{\rm top}(X') = h_{\rm top}\left(\bigcup_{n=1}^\infty X'_{\varepsilon_n}\right) = \sup_n h_{\rm top}(X'_{\varepsilon_n}),$$

therefore $h_{\rm top}(X') \geqslant h$.

We have proved that for each $X'' \subset X$ such that $\mu(X'') = 1$ we have $h_{\rm top}(X'' \cap \mathcal{G}_h) = h(\mu)$, from what we readily derive the desired conclusion. □

A general approach, where the existence of pointwise dimension is not needed, is developed in [39]. However, since we are placed in a framework where the pointwise dimension does exist, then for the sake of clearness, we have chosen to present it not in that generality. However we will briefly discuss this general approach at the end of this section.

12.2. A formula for the dimension for a measure

We place ourselves in the same context as for Theorem 11.9, i.e., (Ω, σ) is a specified subshift supplied with the ultrametric $d_\Omega(\omega, \varpi) = \exp(-u(\zeta^n(\omega)))$, and the Borel probability measure μ is ergodic and such that $h(\mu) > 0$. Let us remind that, for each $q \in \mathbb{R}$

$$\alpha^\mu(q) := \inf\left\{\alpha(q, X')\colon\ X' \subset \Omega \text{ and } \mu(X') = 1\right\},$$

and that

$$\lambda_\mu = \int_\Omega u(\omega)\, d\mu(\omega)$$

is the Lyapunov exponent of the measure μ with respect to the ultrametric d_Ω. We have the following.

THEOREM 12.2. *Let (Ω, σ), μ and d_Ω be as above, then for every $q \in \mathbb{R}$*

$$\alpha^{\mu}(q) = \frac{h(\mu) - q}{\lambda_\mu}.$$

PROOF. Theorem 11.9 ensures that

$$\lim_{n\to\infty} \frac{\log \mu(\zeta^n(\omega)) + q\tau(\zeta^n(\omega))}{\log(\operatorname{diam} \zeta^n(\omega))} = d := \frac{h(\mu) - q}{\lambda_\mu}$$

for all $q \in \mathbb{R}$ and μ-almost all $\omega \in \Omega$. Proceeding as in the previous proof, let $X'' \subset \Omega$ be such that $\mu(X'') = 1$ and define

$$\mathcal{G}_d := \{\omega \in \Omega\colon\ d_{\mu,q}(\omega) = d\}.$$

The set $X' := X'' \cap \mathcal{G}_d$ has full measure μ.

Once again we invoke the Egorov theorem which, together with Theorem 11.9, tells us that for $\varepsilon > 0$ it exists $X'_\varepsilon \subset \Omega$ with $\mu(X'_\varepsilon) \geqslant 1 - \varepsilon$, such that for each $\delta > 0$ there exists $n(\delta) \in \mathbb{N}$ such that

$$e^{-\delta}\mu\big(\zeta^n(\omega)\big) \leqslant \big(\operatorname{diam} \zeta^n(\omega)\big)^d e^{-q\tau(\zeta^n(\omega))} \leqslant e^{\delta}\mu\big(\zeta^n(\omega)\big), \tag{12.2}$$

for all $n \geqslant n(\delta)$ and $\omega \in X'_\varepsilon$. For each $n \in \mathbb{N}$ let $\eta_n := \max_{\omega\in\Omega} e^{-u(\zeta^n(\omega))}$. Then, for each cover $\mathcal{C}$ of X'_ε by balls of diameter smaller than or equal to $\eta_{n(\delta)}$, i.e., by cylinders of length $n \geqslant n(\delta)$, using (12.2) we obtain

$$M(d, q, \mathcal{C}, X'_\varepsilon) := \sum_{B\in\mathcal{C}} \big(\operatorname{diam}(B)\big)^d e^{-\tau(B)} \geqslant e^{-\delta} \sum_{B\in\mathcal{C}} \mu(B) \geqslant e^{-\delta}\mu(X'_\varepsilon).$$

Since this is true for any cover $\mathcal{C}$ of X'_ε by balls of diameter $\eta \leqslant \eta_{n(\delta)}$, and for all $\delta > 0$, then we have

$$\lim_{\eta\to 0^+} \big(\inf\{M(d, q, \mathcal{C}, X'_\varepsilon)\colon\ \operatorname{diam}(\mathcal{C}) \leqslant \eta\}\big) \geqslant 1 - \varepsilon.$$

From this it follows that $\alpha(q, X'_\varepsilon) \geqslant d$. Since $X' \supset X'_\varepsilon$, then $\alpha(q, X') \geqslant d$ too.

For the converse inequality fix $\varepsilon > 0$ and $\delta > 0$ as before, and let $\mathcal{C}_n$ be the cover of X'_ε by cylinders of length $n \geqslant n(\delta)$, which is also a particular cover by balls of diameter smaller than or equal to $\eta_{n(\delta)}$. Since this cover is at the same time a partition, using (12.2) we obtain

$$M(d, q, \mathcal{C}_n, X'_\varepsilon) = \sum_{B\in\mathcal{C}_n} \operatorname{diam}(B)^d e^{-q\tau(B)} \leqslant e^{\delta} \sum_{B\in\mathcal{C}_n} \mu(B) \leqslant e^{\delta},$$

which holds for all $\delta > 0$. Therefore

$$\begin{aligned}\lim_{\eta\to 0^+} \big(\inf\{M(d, q, \mathcal{C}, X'_\varepsilon)\colon\ \operatorname{diam}(\mathcal{C}) \leqslant \eta\}\big) &\leqslant \limsup_{n\to\infty} M(d, q, \mathcal{C}_n, X'_\varepsilon) \\ &\leqslant 1,\end{aligned}$$

which implies the inequality $\alpha(q, X'_\varepsilon) \leqslant d$ for all $\varepsilon > 0$.

Following the lines of the proof of Proposition 12.1, let $\{\varepsilon_n\}_{n=1}^{\infty}$ be any positive sequence converging to zero. Then $X' = \bigcup_{n=1}^{\infty} X'_{\varepsilon_n}$, and

$$\alpha(q, X') = \alpha\left(q, \bigcup_{n=1}^{\infty} X'_{\varepsilon_n}\right) = \sup_n \alpha(q, X'_{\varepsilon_n}),$$

therefore $\alpha(q, X') \leqslant d$.

We have proved that for each $X'' \subset \Omega$ such that $\mu(X'') = 1$ we have $\alpha(q, X'' \cap \mathcal{G}_d) = d$. The conclusion follows. □

Thus, Theorems 12.2 and 11.9 imply that $\alpha^{\mu}(q) = d_{\mu,q}(\omega)$ for all ω in a set of full measure μ.

12.3. The q-pointwise dimension for suspended flows

Theorem 12.2 has a direct extension for suspended flows over the specified subshift (Ω, σ). To show it, we have to start with some definitions. For the suspended flow (Ω^{ϕ}, Φ), consider a Φ-invariant ergodic probability measure $\bar{\mu}$. Fubini's theorem allows us to decompose $\bar{\mu}$ into a "horizontal projection" μ, which is a σ-invariant ergodic measure over Ω, and a "vertical projection" equivalent to the Lebesgue measure on each fiber. Hence, $\bar{\mu}$ is such that the equality

$$\int_{\Omega^{\phi}} F\, d\bar{\mu} := \frac{\int_{\omega}(\int_0^{\phi(\omega)} F(\omega, t)\, dt)\, d\mu(\omega)}{\int_{\Omega} \phi(\omega)\, d\mu(\omega)}$$

holds.

For a Φ-invariant probability measure $\bar{\mu}$, and for $q \in \mathbb{R}$, let

$$\alpha^{\bar{\mu}}(q) := \inf\{\alpha(Y, q)\colon\ Y \subset \Omega^{\phi},\ \bar{\mu}(Y) = 1\},$$

with $\alpha(Y, q)$ such as in Chapter 10. In what follows we will consider, in the definition of $\alpha(Y, q)$, covers of Y by squared rectangles $R(x_s, \varepsilon, \varepsilon)$. In Chapter 10 we have proved that this spectrum coincides with the one defined by using covers by rectangles $R(x_s, \varepsilon, \delta)$, as long as $\alpha - 1$ and q are non-negative.

The analogous of Theorem 12.2 for suspended flows stands as follows.

THEOREM 12.3. *For $q \in \mathbb{R}$ and for $\bar{\mu}$ a Φ-invariant ergodic probability measure such that $h(\mu) := h(\mu|\sigma) > 0$, the spectrum for Poincaré recurrences $\alpha^{\bar{\mu}}(q)$ satisfies the equation*

$$h(\mu) + \int_{\Omega} \left((1 - \alpha^{\bar{\mu}})u(\omega) - q\phi(\omega)\right) d\mu(\omega) = 0.$$

Let us remind that μ is the horizontal projection of the Φ-invariant measure $\bar{\mu}$, which is a σ-invariant probability measure on Ω, and the Hölder continuous function $u:\Omega\to\mathbb{R}$ is the function defining the ultrametric d_Ω in Ω.

The proof of Theorem 12.3 relies on three elements we already have at our disposal:

(a) Theorem 11.3, which establishes that $\lim_{n\to\infty}\tau(\zeta^n(\omega))/n=1$, for μ-almost every $\omega\in\Omega$.
(b) The SMB-Theorem.
(c) Theorem 12.2, whose proof can readily be adapted to derive the equality

$$\lim_{\varepsilon\to 0}\frac{\bar{\mu}(R(x_s,\varepsilon,\varepsilon))+q\tau_\Phi(R(x_s,\varepsilon,\varepsilon))}{\log(\varepsilon)}=\alpha^{\bar{\mu}}(q)$$

for $\bar{\mu}$ almost every x_s.

PROOF OF THEOREM 12.3.

Step 1. First we prove that the limit

$$d_{\bar{\mu},q}:=\lim_{\varepsilon\to 0}\frac{\log(\bar{\mu}(R(x_s,\varepsilon,\varepsilon)))+q\tau_\Phi(R(x_s,\varepsilon,\varepsilon))}{\log(\varepsilon)}$$

exits, and is constant in a set of full $\bar{\mu}$ measure.

As we proved in Chapter 10, the horizontal projection $R_X(x_s,\varepsilon)$ of the open rectangle $R(x_s,\varepsilon,\varepsilon)$ coincides with a cylinder $\zeta^{m_{x_s,\varepsilon}}(\omega)$ with

$$m_{x_s,\varepsilon}:=\min\bigl\{n\in\mathbb{N}\colon\ (1-s)e^{-u(\zeta^n(\omega))}+se^{-u(\zeta^{n-1}(\sigma\omega))}<\varepsilon\bigr\}.$$

Since ϕ is a Hölder continuous function, there is a constants $C_0\geqslant 1$ such that

$$\begin{aligned}\bar{\mu}\bigl(R(x_s,\varepsilon,\varepsilon)\bigr)&:=\varepsilon\,\frac{\int_{\zeta^{m(x_s,\varepsilon)}(\omega)}\phi(\varpi)\,d\mu(\varpi)}{\int_\Omega\phi(\varpi)\,d\mu(\varpi)}\\&=\varepsilon\,\mu\bigl(\zeta^{m(x_s,\varepsilon)}(\omega)\bigr)\,\frac{C_0^{\pm 1}\phi(\omega)}{\int_\Omega\phi(\varpi)\,d\mu(\varpi)}.\end{aligned}$$

Here $A=C_0^{\pm 1}B$ is a shortcut for $C_0^{-1}B\leqslant A\leqslant C_0B$. Thus, the SBM theorem, ensures that

$$\lim_{\varepsilon\to 0}\frac{\log(\bar{\mu}(R(x_s,\varepsilon,\varepsilon)))}{m(x_s,\varepsilon)}=-h(\mu)+\lim_{\varepsilon\to 0}\frac{\log(\varepsilon)}{m(x_s,\varepsilon)},$$

whenever the last limit exist.

Since $u:\Omega\to\mathbb{R}$ is also a Hölder continuous function, and because of the definition of $m(x_s,\varepsilon)$ given above, there is a constant $C_1\geqslant 1$ such that $\varepsilon=C_1^{\pm 1}\exp(-u(\zeta^{m(x_s,\varepsilon)}(\omega)))$, therefore

$$\log\varepsilon=\sum_{j=0}^{m(x_s,\varepsilon)}u\bigl(T^j(\omega)\bigr)\pm C$$

for a positive constant C which depends on $\log(C_1)$ and u.

As we proved in Chapter 10, there is a positive constant C_2 such that

$$\tau_\Phi\big(R(x_s,\varepsilon)\big) = \sum_{j=0}^{m_{x_s,\varepsilon}-1} \phi\big(\sigma^j\omega\big) \pm C_2.$$

From all this, and taking into account Theorem 11.3 and Birkhoff's ergodic theorem we obtain

$$\lim_{\varepsilon\to 0} \frac{\log(\bar{\mu}(R(x_s,\varepsilon,\varepsilon))) + q\tau_\Phi(R(x_s,\varepsilon,\varepsilon))}{\log(\varepsilon)}$$
$$= d := 1 + \frac{\int_\Omega \phi(\omega)\,d\mu(\omega) - h(\mu)}{\int_\Omega u(\omega)\,d\mu(\omega)}$$

for $\bar{\mu}$ almost every $x_s \in \Omega^\phi$.

Step 2. Now we prove that the $\bar{\mu}$ almost everywhere constant value of the previous limit coincides with the dimension of a measure $\alpha^{\bar{\mu}}(q)$. We proceed as in the proofs of Theorem 12.2 above. Let $Z \subset \Omega^\phi$ be such that $\bar{\mu}(Z) = 1$ and define

$$\mathcal{G}_d := \big\{\omega \in \Omega\colon\ d_{\bar{\mu},q}(\omega) = d\big\}.$$

The set $Y := Z \cap \mathcal{G}_d$ has full measure $\bar{\mu}$. Let $\{\varepsilon_n\}_{n=1}^\infty$ be a positive sequence converging to zero. The Egorov theorem ensures that for $\eta > 0$ there exists $Y_\eta \subset \Omega^\phi$ with $\bar{\mu}(Y_\eta) \geqslant 1-\eta$ such that for each $\delta > 0$ there exists $n(\delta) \in \mathbb{N}$ such that

$$(\varepsilon_n)^d e^{-q\tau_\Phi(R(x_s,\varepsilon_n,\varepsilon_n))} = e^{\pm\delta}\bar{\mu}\big(R(x_s,\varepsilon_n,\varepsilon_n)\big)$$

for all $n \geqslant n(\delta)$ and $x_s \in Y_\eta$. Then, for each cover $\mathcal{R}$ of Y_η by squared rectangles $R(x_s,\varepsilon_n,\varepsilon_n)$, with $n = n(\delta)$, we have

$$M(d,q,\mathcal{R},Y_\eta) := \sum_{R\in\mathcal{R}} \big(\operatorname{diam}(R)\big)^d e^{-\tau_\Phi(R)}$$
$$\geqslant Ce^{-\delta}\sum_{R\in\mathcal{R}} \bar{\mu}(R) \geqslant Ce^{-\delta}\mu(Y_\eta),$$

where $C = 2^d$ or $C = 2^{-d}$, depending on the sign of d. Since this is true for any cover $\mathcal{R}$ of Y_η by squared rectangles $R(x_s,\varepsilon_n,\varepsilon_n)$, with $n \geqslant n(\delta)$, then

$$\lim_{\varepsilon\to 0^+}\big(\inf\big\{M(d,q,\mathcal{R},Y_\eta)\colon\ \operatorname{diam}(\mathcal{R}) \leqslant \varepsilon\big\}\big) \geqslant 1-\eta.$$

Here the infimum is taken over all covers by squared rectangles of diameter smaller than or equal to ε. Hence, we have $\alpha(q,Y_\eta) \geqslant d$, and since $Y \supset Y_\eta$, then $\alpha(q,Y) \geqslant d$ too.

For the converse inequality fix $\eta > 0$ and $\delta > 0$ as before. For each $n \in \mathbb{N}$ let $\mathcal{R}_n$ be a minimal cover of Y_η by squared rectangles $R(x_s,\varepsilon_n,\varepsilon_n)$, all with the same ε_n. For $n \geqslant n(\delta)$ we have

$$M(d, q, \mathcal{R}_n, Y_\eta) = \sum_{R \in \mathcal{R}_n} \operatorname{diam}(R)^d e^{-q\tau(B)}$$
$$\leqslant C^{-1} e^{\delta} \sum_{R \in \mathcal{R}_n} \bar{\mu}(R) \leqslant 2C^{-1} e^{\delta},$$

with C as above. The factor 2 is added because $\mathcal{R}_n$ is not a partition, but an open cover. Nevertheless, since it is minimal, we can avoid horizontal overlaps, and we only would have vertical overlaps which are one-dimensional.

Now, since this can be done for all $\delta > 0$, then

$$\lim_{\varepsilon \to 0^+} \left(\inf\left\{M(d, q, \mathcal{R}, Y_\eta) : \operatorname{diam}(\mathcal{R}) \leqslant \varepsilon\right\}\right)$$
$$\leqslant \limsup_{n \to \infty} M(d, q, \mathcal{R}_n, Y_\eta) \leqslant 2C^{-1},$$

which implies $\alpha(q, Y_\eta) \leqslant d$ for all $\eta > 0$.

Following the lines of the proof of Proposition 12.1, let $\{\eta_m\}_{n=1}^{\infty}$ be any positive sequence converging to zero, then $Y = \bigcup_{m=1}^{\infty} Y_{\eta_m}$, and

$$\alpha(q, Y) = \alpha\left(q, \bigcup_{m=1}^{\infty} Y_{\eta_m}\right) = \sup_m \alpha(q, Y_{\eta_m}),$$

therefore $\alpha(q, Y) \leqslant d$.

We have proved that for each $Z \subset X^{\phi}$ such that $\mu(Z) = 1$ we have $\alpha(q, Z \cap \mathcal{G}_d) = d$.

Last step. From the parts 1 and 2 of this proof it follows that

$$1 + \frac{\int_\Omega \phi(\omega)\, d\mu(\omega) - h(\mu)}{\int_\Omega u(\omega)\, d\mu(\omega)} = \alpha^{\bar{\mu}}(q),$$

which we can write also as

$$h(\mu) + \int_\Omega \left(\left(\alpha^{\bar{\mu}}(q) - 1\right) u(\omega) - q\phi(\omega)\right) d\mu(\omega) = 0.$$

□

12.4. Multifractal decomposition for sticky sets

In this section we consider some properties of an invariant measure for a polysymbolic system in the case when the system is minimal. Let us recall first that, the unique measure μ_{q_*} is defined on each cylinder $[\omega_0, \ldots, \omega_{L-1}]$ by

$$\mu_{q_*} = \frac{1}{q_0 \cdots q_{L-1}}. \tag{12.3}$$

Because of Proposition 2.2, the system $f|F$ is uniquely ergodic either, and the corresponding measure is the pushed forward one by the conjugacy h. Denote the measure by ν_{q*}.

From now on we restrict ourselves to the case of the full shift $\Omega_{q*} = \Omega_2 := \{0, 1\}^{\mathbb{N}}$. We assume that our sticky set is of the planetary type (see Section 3.5.1) and that there exist numbers $0 < \lambda_0 < \lambda_1 < 1$ and constants $\underline{c}, \bar{c} > 0$ such that

$$\underline{c} \prod_{j=0}^{n-1} \lambda_{\omega_j} \leqslant \operatorname{diam} \Delta_{\underline{\omega}} \leqslant \bar{c} \prod_{j=0}^{n-1} \lambda_{\omega_j} \tag{12.4}$$

where $\Delta_{\underline{\omega}} \supset \chi(\underline{\omega})$, $\underline{\omega} = [\omega_0, \dots, \omega_{n-1}]$. As it was shown in Section 3.5.1, the Hausdorff dimension of the sticky set F is the root α_0 of the Moran's equation

$$\lambda_0^{\alpha} + \lambda_1^{\alpha} = 1. \tag{12.5}$$

The measure ν_{q*} in our case is just the Bernoulli measure, say, ν: $\nu(h^{-1}(\underline{\omega})) = 2^{-n}$ if $\underline{\omega} = [\omega_0, \dots, \omega_{n-1}]$. If we take into account that the Bernoulli measure is an invariant with respect to the shift map on Ω_2, and remember that the sticky set F is resulting from a Moran geometric construction (see Section 3.5.1), then we may forget about dynamics on F and just make multifractal analysis of the measure ν in the same way as people do it for conformal repellers (see, for instance, [97] and [100]). In order to apply these results, we need to show that the measure ν is not of full dimension. For that, we need to construct a set of arbitrary large measure with "small" Hausdorff dimension. The idea is simple enough. The density of appearance of zeros and ones in a typical sequence of symbols should be $1/2$. So, a desired set should be constructed by cylinders containing approximately equal numbers of zeros and ones. We prove the following.

THEOREM 12.4. *Whenever $\lambda_0 \neq \lambda_1$, the uniform (Bernoulli) measure ν on F is not of full Hausdorff dimension.*

PROOF. We prove, first, that the measure μ is not of full Hausdorff dimension, provided that Ω_2 is endowed with the metric d_Ω. The proof consists of three lemmas. In Lemma 12.1 below, sets of arbitrary large measure are constructed. In Lemma 12.2, an upper bound $\bar{\alpha}$ for their Hausdorff dimension α_μ is computed. In Lemma 12.3, we prove that the upper bound $\bar{\alpha}$ is smaller than the Hausdorff dimension α of the set Ω_2, if $\lambda_0 \neq \lambda_1$.

For $L \in \mathbb{Z}^+$, consider the sets $B(\delta, L) = \{\omega \in \Omega_2\colon |n_0(\omega[L]) - n_1(\omega[L])| \leqslant \delta L\}$ where $\omega[L] = \omega_0 \dots \omega_{L-1}$ and $n_i(\omega[L]) = \#\{\omega_j = i\colon j = 0, \dots, L-1\}$, $i = 0, 1$. We are interested in the sets $C(\delta, L_0) = \bigcap_{L \geqslant L_0} B(\delta, L)$, for $L_0 \gg 1$ to get an estimate of the Hausdorff dimension of μ [97],

$$\dim_H \mu = \lim_{\delta \to 0} \inf\bigl\{\dim_H Z\colon \mu(Z) \geqslant 1 - \delta\bigr\}.$$

We will choose the sets $C(\delta, L)$ in the capacity of the sets Z in the formula above.

Every set $B(\delta, L)$ is a union of cylinders defined by the words of length L in the set $W(\delta, L) = \{\omega[L] \in \{0,1\}^L\colon |n_0(\omega[L]) - n_1(\omega[L])| \leqslant \delta L\}$. The measure of each set is $\mu(B(\delta, L)) = N_L/2^L$, where $N_L = \#W(\delta, L)$. To estimate the value of N_L for $L \gg 1$ we use Stirling's approximation as follows

$$\begin{aligned} N_L &= \sum_{k=\lceil L\frac{1-\delta}{2}\rceil}^{\lfloor L\frac{1+\delta}{2}\rfloor} \binom{L}{k} = 2^L - 2\sum_{k\leqslant L\frac{1-\delta}{2}} \binom{L}{k} \\ &\approx 2^L\left(1 - K_\delta 2^{-L(1-H((1-\delta)/2))-(1/2)\ln L}\right) \end{aligned} \tag{12.6}$$

with K_δ a positive number and $H(\alpha) := -\alpha\ln\alpha - (1-\alpha)\ln(1-\alpha)$. Hence, we are ready to prove the following result.

LEMMA 12.1. *For arbitrary $\delta > 0$ and $L_0 \gg 1$ there exist $K > 0$ and $K'_\delta > 0$ such that $\mu(C(\delta, L_0)) \geqslant 1 - K'_\delta\, 2^{-KL_0}$.*

PROOF.

$$\mu\big(C(\delta, L_0)\big) = \mu\Big(\bigcap_{L\geqslant L_0} B(\delta, L)\Big) \geqslant 1 - \sum_{L\geqslant L_0} \mu\big(B(\delta, L)^c\big)$$

where

$$\begin{aligned} \sum_{L\geqslant L_0} \mu\big(B(\delta, L)^c\big) &= \sum_{L\geqslant L_0}\left(1 - \frac{N_L}{2^L}\right) \approx K_\delta \sum_{L\geqslant L_0} 2^{-LK-(1/2)\ln L} \\ &\leqslant \frac{K_\delta}{1-2^{-K}}\, 2^{-L_0K} \end{aligned}$$

with $K = 1 - H((1-\delta)/2) > 0$ for sufficiently small $\delta > 0$. □

LEMMA 12.2.

$$\dim_H C(\delta, L_0) \leqslant \frac{\ln 2}{-\frac{1}{2}(\ln\lambda_0 + \ln\lambda_1)} =: \bar{\alpha}.$$

PROOF. To obtain an estimate from above, we can choose an arbitrary family of covers of the set $C(\delta, L_0)$. Let us consider the sets $B(\delta, L)$, $L \geqslant L_0$, in the capacity of such a family. Consider the sum

$$\sum_{\omega_0,\ldots,\omega_{L-1}} \left(\prod_{k=0}^{L-1} \lambda_{\omega_k}\right)^\alpha \tag{12.7}$$

with $(\omega_0, \dots, \omega_{L-1}) \in W(\delta, L)$. Remark that $[\omega_0, \dots, \omega_{L-1}] \subset B(\delta, L)$ for $(\omega_0, \dots, \omega_{L-1}) \in W(\delta, L)$ and that $B(\delta, L)$, for $L \geqslant L_0$, covers $C(\delta, L)$. Assume that $\lambda_1 \geqslant \lambda_0$, then

$$\prod_{k=0}^{L-1} \lambda_{\omega_k} = (\lambda_0\lambda_1)^{L/2}\left(\frac{\lambda_0}{\lambda_1}\right)^{n_0^{(L)}(\omega)-L/2} \leqslant (\lambda_0\lambda_1)^{L/2}\left(\frac{\lambda_1}{\lambda_0}\right)^{\delta L/2}.$$

Thus

$$\sum_{\omega_0,\dots,\omega_{L-1}} \left(\prod_{k=0}^{L-1} \lambda_{\omega_k}\right)^{\alpha} \leqslant \left[(\lambda_0\lambda_1)^{L/2}\left(\frac{\lambda_1}{\lambda_0}\right)^{\delta L/2}\right]^{\alpha} \sum_{k=L\frac{1-\delta}{2}}^{L\frac{1+\delta}{2}} \binom{L}{k}$$
$$\leqslant \left[2(\lambda_0\lambda_1)^{\alpha/2}\left(\frac{\lambda_1}{\lambda_0}\right)^{\delta\alpha/2}\right]^{L}.$$

This expression remains finite as L goes to infinity for those values of α for which $2(\lambda_0\lambda_1)^{\alpha/2} \cdot (\lambda_1/\lambda_0)^{\alpha\delta/2} \leqslant 1$. It implies the validity of the inequality $2(\lambda_0\lambda_1)^{\alpha/2} \leqslant 1$ and proves the lemma. □

Denote by α_μ the dimension of the measure μ.

COROLLARY 12.1. $\alpha_\mu < \bar{\alpha}$.

The proof follows directly from Lemma 12.2 and equality (3.3) of the book [97], p. 22.

LEMMA 12.3. *For $\lambda_0 \neq \lambda_1$, $\alpha_\mu < \alpha$, the Hausdorff dimension of the set Ω_2.*

PROOF. From (12.5) and Lemma 12.2, it follows that for every positive γ,

$$\alpha(\lambda_0, \lambda_1) = \gamma\alpha\left(\lambda_0^\gamma, \lambda_1^\gamma\right) \quad \text{and} \quad \bar{\alpha}(\lambda_0, \lambda_1) = \gamma\bar{\alpha}\left(\lambda_0^\gamma, \lambda_1^\gamma\right). \tag{12.8}$$

Assume, first, that $\lambda_1 = 1-\lambda_0$. Then, by Moran's equation (12.5), $\alpha(\lambda_0, 1-\lambda_0) = 1$, and

$$\bar{\alpha}(\lambda_0, 1-\lambda_0) < \max_{\lambda_0} \frac{2\ln 2}{-\ln\lambda_0(1-\lambda_0)} = 1 \tag{12.9}$$

if $\lambda_0 \neq 1/2$. For arbitrary λ_0' and λ_1', due to property (12.8), there exist γ and λ_0 such that $\lambda_0' = \lambda_0^\gamma$, $\lambda_1' = (1-\lambda_0)^\gamma$, $\alpha(\lambda_0', \lambda_1') = \gamma^{-1}$, and $\bar{\alpha}(\lambda_0', \lambda_1') = \gamma^{-1}\bar{\alpha}(\lambda_0, 1-\lambda_0)$. Hence, after the inequality (12.9) and Corollary 12.1, it follows that $\alpha_\mu < \bar{\alpha}(\lambda_0', \lambda_1') < \alpha(\lambda_0', \lambda_1')$, for arbitrary $\lambda_0' \neq \lambda_1'$. □

To conclude the proof of Theorem 12.4, we apply Proposition 4.1 and Theorem 6.3 of the book [97], which tells us that the Hausdorff dimension is an invariant with respect to a Lipschitz continuous homeomorphism with a Lipschitz continuous inverse. □

REMARK 12.1. For the Markov map of the interval g of Example 3.1 the g-invariant measure $\tilde{\nu}$ is not of full dimension either. The map g has a conformal repeller J that is the support of the measure $\tilde{\nu}$, and all assumptions of Theorem 21.1 in the book [97] are satisfied. Thus, it has the multifractal decomposition described in this theorem.

If the geometric constructions of the sets J and F are modeled by the same symbolic system then they are metrically equivalent (see Proposition 12.2 below). Therefore, the multifractal decompositions of $\tilde{\nu}|J$ and $\nu|F$ are the same. This means the following. Assume that the limit

$$d_\nu(x) := \lim_{r\to 0} \frac{\log \nu(B(x,r))}{\log r}$$

exists. It is called the pointwise dimension at the point $x \in F$ [124]. Let $F_\alpha = \{x \in F\colon d_\nu(x) = \alpha\}$ and the Hausdorff dimension $\dim_H F_\alpha =: f_\nu(\alpha)$, then the function $f_\nu(\alpha)$ is said to be the multifractal spectrum of ν (see, for instance, [97]). The representation of F in the form $F = \bigcup_\alpha F_\alpha \cup \widehat{F}$, where the irregular part $\widehat{F}$ is the set of points for which the pointwise dimension does not exist, is called the multifractal decomposition of the set F. Now, let

$$T_\nu(q) = -\lim_{r\to 0} \frac{\log \inf_{G_r} \sum_{B\in G_r} \nu(B)^q}{\log r}$$

where the infimum is taken over all finite covers G_r of F by open balls of radius r. Remark that

$$T_\nu(q) = (1-q)HP_q(\nu)$$

where $HP_q(\nu)$ is the Hentschel–Procaccia spectrum (provided that corresponding limits exist, see, for instance, [97]).

The following result holds.

THEOREM 12.5.

(i) $T_\nu(q) = T_{\tilde{\nu}}(q)$.
(ii) $f_\nu(\alpha) = f_{\tilde{\nu}}(\alpha)$, *and* $F_\alpha = \chi(\tilde{\chi}^{-1}(J_\alpha))$ *where* $J_\alpha = \{x \in J\colon d_{\tilde{\nu}}(x) = \alpha\}$.

PROOF. Since the map $\chi \circ \tilde{\chi}^{-1}$ is Lipschitz continuous with a Lipschitz continuous inverse, then the statement (i) follows from Theorem 8.3 (p. 50), equality (18.1) (p. 182), and Statement 5 of Theorem 21.1 (p. 214) of the book [97].

Proof of the statement (ii). Let $x \in J_\alpha$, i.e., $\lim_{r\to 0} \log \tilde{\nu}(B(x,r))/\log r = \alpha$. Let L (correspondingly l) be a Lipschitz constant of the map $\chi \circ \tilde{\chi}^{-1}$ (correspondingly $\tilde{\chi} \circ \chi^{-1}$) and $y = \chi(\tilde{\chi}^{-1}(x))$. Then

$$B(y, l^{-1}r) \subset \chi\bigl(\tilde{\chi}^{-1}\bigl(B(x,r)\bigr)\bigr) \subset B(y, Lr)$$

and

$$\nu\bigl(B\bigl(y, l^{-1}r\bigr)\bigr) \leqslant \nu\bigl(\chi\bigl(\tilde{\chi}^{-1}\bigl(B(x,r)\bigr)\bigr)\bigr) = \tilde{\nu}\bigl(B(x,r)\bigr) \leqslant \nu\bigl(B(y, Lr)\bigr).$$

Therefore, $\alpha = d_\nu(y)$. In the same way we show that if $y \in F_\alpha$ then $\tilde{\chi}(\chi^{-1}(y)) \in J_\alpha$. This implies (together with Theorem 6.3 in [97]) the statement (ii). □

COROLLARY 12.2. *The theorem, Theorem* 11.1, *and Theorem* 21.1 *in* [97] *imply that:*

(i) *The pointwise dimension $d_\nu(x)$ exists for ν-almost every $x \in F$ and*

$$d_\nu(x) = \frac{2\log 2}{-(\log \lambda_0 + \log \lambda_1)}.$$

(ii) *The function $f_\nu(\alpha)$ is defined on the interval $[\alpha_1, \alpha_2]$ which is the range of the function $\alpha(q)$ (i.e., $0 \leqslant \alpha_1 \leqslant \alpha_2 < \infty$, $\alpha_1 = \alpha(\infty)$ and $\alpha_2 = \alpha(-\infty)$); this function is real analytic and $f_\nu(\alpha(q)) = T(q) + q\alpha(q)$.*

(iii) *The functions $f_\nu(\alpha)$ and $T_\nu(q)$ are strictly convex and form a Legendre transform pair.*

REMARK 12.2. The results of this section tell us that a minimal multipermutative system, being uniquely ergodic, nevertheless possesses a nontrivial multifractal decomposition, provided that the rates of contraction are different. If rates of contraction are the same, then multifractal decomposition is trivial (i.e., has the only element plus the irregular part). This follows from Proposition 12.2 below and [97].

12.5. Remarks on the q-pointwise dimension and the dimension of a measure

REMARK 12.3 *(Dimension of measures).* Now come back to Section 4.1 and assume that μ is a Borel probability measure supported on the set $Z \subset \mathbb{R}^m$. The quantity

$$\dim_c \mu := \inf\bigl\{\alpha_c(Y)\colon\ \mu(Y) = 1\bigr\}$$

is said to be the Carathéodory dimension of the measure μ [97]. In particular, $\dim_H \mu = \inf\{\dim_H Y\colon\ \mu(Y) = 1\}$ is called the Hausdorff dimension of the measure μ.

In the case where $d_\mu(x)$ exists, the quantity $d_H(\mu)$ can be estimated. Theorem 7.1 in [97] tells us that:

(i) if $\underline{d}_\mu(x) \geqslant d$ for μ-a.e. x then $\dim_H \mu \geqslant d$;
(ii) if $\overline{d}_\mu(x) \leqslant d$ for μ-a.e. x then $\dim_H \mu \leqslant d$;
(iii) if $d_\mu(x) = d$ for μ-a.e. x then $\dim_H \mu = d$.

It is clear that $\dim_c \mu \leqslant \alpha_c(Z)$ for any Carathéodory structure and any measure μ. *A measure* μ_0 *is said to be the measure of full dimension if* $\dim_c \mu_0 = \alpha_c(Z)$. The measure of full dimension may not exist. For example, even for two-dimensional Axiom-A diffeomorphisms, the measure of full dimension almost never exists [82]. Nevertheless, there are interesting situations where it is so.

EXAMPLE 12.1 *(Existence of the measure of full dimension).* Let us consider the conformal repeller J in Example 3.1. We know that the Hausdorff dimension $\alpha_c = \dim_H J$ is the root of the Moran equation

$$\lambda_0^{\alpha_c} + \lambda_1^{\alpha_c} = 1.$$

For the sake of definiteness, assume that $\lambda_0 < \lambda_1$ and let $\kappa := \log \lambda_1 / \log \lambda_0$, i.e., $\lambda_1 = \lambda_0^\kappa$. Introduce a number $p > 0$ satisfying the equation

$$p + p^\kappa = 1.$$

We show that $p = \lambda_0^{\alpha_c}$. Indeed,

$$\lambda_0^{\alpha_c} + \left(\lambda_0^{\alpha_c}\right)^\kappa \equiv \lambda_0^{\alpha_c} + \lambda_1^{\alpha_c} = 1,$$

by definition of α_c. Consider the $(p, 1-p)$-Bernoulli measure m on Ω_2 (in other words, the measure of the cylinder $m([\omega_0, \dots, \omega_{n-1}]) = \prod_{k=0}^{n-1} p_{\omega_k}$ where $p_{\omega_k} = p$ if $\omega_k = 0$, $p_{\omega_k} = p^\kappa = 1-p$ if $\omega_k = 1$). Denote by μ the pushed-forward measure on J and show that μ is the measure of full dimension. We have $h_\mu(f) = p \log p + (1-p)\log(1-p)$, $\chi_\mu^+ = p \log \lambda_0 + (1-p) \log \lambda_1$. Furthermore, for the conformal repeller J one has

$$d_\mu(x) = \frac{h_\mu(f)}{\chi_\mu^+},$$

where χ_μ^+ is the Lyapunov exponent with respect to measure μ (the proof is a simplified version of the result in [124] – see [97]). Hence,

$$\begin{aligned} d_\mu(x) &= \frac{p \log p + \kappa p^\kappa \log p}{p \log \lambda_0 + p^\kappa \log \lambda_1} \\ &= \log p \frac{1 + p^{\kappa-1} \log \lambda_1 / \log \lambda_0}{\log \lambda_0 + p^{\kappa-1} \log \lambda_1} \\ &= \frac{\log p}{\log \lambda_0} \end{aligned}$$

for μ-a.a. x. Because of the mentioned Theorem 7.1 in [97], we have that

$$\dim_H \mu = \frac{\log p}{\log \lambda_0} = \frac{\alpha_c \log \lambda_0}{\log \lambda_0} = \alpha_c.$$

This example is a particular manifestation of the general result [100] that says that if m is an equilibrium measure on a subshift for the potential $(\omega_0, \omega_1, \ldots) \mapsto \alpha_c \log \lambda_{\omega_0}$ and μ is pushed-forward on F, the result of the corresponding Moran type construction with $\dim_H F = \alpha_c$, then μ is a measure of full dimension and

$$\alpha_c = -\frac{h_\mu}{\int_\Omega \log \lambda_{\omega_0}\, dm}. \tag{12.10}$$

A similar theorem (Theorem 20.1 in [97]) holds for conformal repellers.

REMARK 12.4. In [39] the authors derive a relationship between a lower q-pointwise dimension of the measure μ, and the spectrum of the measure. In that paper the authors treat the problem in a general context, and their main motivation was to propose a local quantity always well defined (a lower limit in this case), from which one could derive the spectrum of dimensions for a measure. This approach applies to the spectra of dimensions of monotone functions of sets other than Poincaré recurrences, but it is particularly adapted to the spectrum of Poincaré recurrences.

The general context is the following: $X \subset \mathbb{R}^n$ is a Borel set, μ is a Borel probability measure, and ψ is a real-valued set function. Then, for $\alpha, q \in \mathbb{R}$ and $Y \subset X$ let

$$M_\psi(\alpha, q, Y) := \lim_{\varepsilon \to 0^+} \inf_{\mathcal{B}\colon \operatorname{diam}(\mathcal{B}) \leqslant \varepsilon} \left\{ \sum_{B \in \mathcal{B}} e^{-q\psi(B)} (\operatorname{diam} B)^\alpha \right\},$$

where the infimum is taken over all finite or countable covers of Y by open balls of diameter smaller or equal ε. The general theory of Carathéodory dimensions ensures, for each $q \in \mathbb{R}$, the existence of a unique number

$$\alpha_\psi(q, Y) := \inf\{\alpha \in \mathbb{R}\colon\ M_\psi(\alpha, q, Y) = \infty\}.$$

The relation $q \to \alpha_\psi(q, Y)$ defines the spectrum of dimensions for ψ of the set Y. Finally,

$$\alpha_\psi^\mu(q) := \inf\{\alpha_\psi(q, Y)\colon\ \mu(Y) = 1\}$$

is the spectrum for ψ of the measure μ.

Here it comes the main difference between this approach and the more traditional approach one can find [97]. Let $Q \subset [0, 1]$ be a countable set such that $0 \in \operatorname{clos} Q$. The lower q-pointwise dimension of μ at $x \in X$ is

$$d_{\mu,q}^\psi(x) := \liminf_{Q \ni \varepsilon \to 0^+} \inf_{y \in B(x,\varepsilon)} \frac{\log \mu(B(y, \varepsilon)) + q\psi(B(y, \varepsilon))}{\log(\varepsilon)}.$$

This limit depends on Q, but for a well behaved function set function ψ, a countable set may be sufficient, i.e., $\liminf_{\varepsilon\to 0^+}$ and $\liminf_{Q\ni\varepsilon\to 0^+}$ coincide. For instance, if ψ is monotonous with respect to the partial order $B \preceq B' \Leftrightarrow B \subset B'$, then $Q := \{1/n\colon\ n \in \mathbb{N}\}$ is sufficient.

The result in [39] is the following.

THEOREM 12.6. *For all $q \in \mathbb{R}$, $\alpha_{\psi}^{\mu}(q) = \text{ess-sup}\, d_{\mu,q}^{\psi}$.*

Let us remind the $\text{ess-sup}\, g := \inf\{a\colon\ \mu\{g(x) < a\} = 1\}$. Theorem 12.6 applies to our case, and it allows us to relate the spectrum for Poincaré dimensions of the measure to the corresponding lower q-pointwise dimension, in the general case when the rate of return times does not exist.

REMARK 12.5. The symbolic setting we consider in this chapter is in perfect agreement with a generalized Moran construction satisfying the gap conditions (3.25) and (3.26). In that case we have the following.

PROPOSITION 12.2. *Let $\chi : X \to F$ be the coding map associated to the generalized Moran construction with model (Ω, σ). Let the construction satisfy the gap conditions* (3.25) *and* (3.26). *Let d_Ω be the ultrametric in X defined as in* (11.1) *by the function $u(\omega) = -\log\lambda(\omega)$. Then, there exist constants $0 < \underline{d} \leqslant \bar{d}$ such that*

$$\underline{d}\ \text{dist}\big(\chi(\omega), \chi(\varpi)\big) \leqslant d_\Omega(\omega, \varpi) \leqslant \bar{d}\ \text{dist}\big(\chi(\omega), \chi(\varpi)\big),$$

for all ω, $\varpi \in \Omega$.

PROOF. Since u is a Hölder continuous function, there exists $C_0 > 1$ such that, for $\zeta^n(\omega)$,

$$\sup_{\varpi'\in\zeta^n(\omega)} \prod_{i=0}^{n-1} \lambda\big(\sigma^i\varpi'\big) \leqslant C_0 \inf_{\varpi'\in\zeta^n(\omega)} \prod_{i=0}^{n-1} \lambda\big(\sigma^i\varpi'\big),$$

uniformly in n and ω. Therefore

$$\begin{aligned} \text{dist}(x, x') &\leqslant \text{diam}\,(\Delta_{\omega_0\cdots\omega_n}) \\ &\leqslant \bar{c} \sup_{\varpi'\in\zeta^n(\omega)} \prod_{i=0}^{n-1} \lambda\big(\sigma^i\varpi'\big) \\ &\leqslant (C_0\,\bar{c})\, e^{-u(\zeta^n(\omega))} \equiv (C_0\,\bar{c}) d_\Omega(\omega, \varpi), \end{aligned}$$

where $x = \chi(\omega)$, $x' = \chi(\varpi)$ and $\varpi \in \zeta^n(\omega)$. If the gap conditions are satisfied, there is a constant (gap) $G > 0$ such that for all $\omega \in \Omega$ and $\varpi \in \zeta^n(\omega) \setminus \zeta^{n+1}(\omega)$

we have

$$\begin{aligned}\operatorname{dist}(x, x') &\geqslant \operatorname{dist}(\Delta_{\omega_0\cdots\omega_n}, \Delta_{\varpi_0\ldots\varpi_n})\\ &\geqslant G \operatorname{diam}(\Delta_{\omega_0\ldots\omega_{n-1}})\\ &\geqslant G\left(\underline{c} \inf_{\varpi'\in\zeta^n(\omega)} \prod_{i=0}^{n-1} \lambda(\sigma^i\varpi')\right)\\ &\equiv (G\,\underline{c})\, e^{-u(\zeta^n(\omega))} \equiv (G\,\underline{c}) d_\Omega(\omega, \varpi).\end{aligned}$$

Hence, the proposition holds when the gap conditions are satisfied. □

Let $f : F \to F$ be such that $f \circ \chi = \chi \circ T$ and ν a f-ergodic Borel probability measure. The equivalence of distance dist in F and distance d_Ω in Ω implies the following.

THEOREM 12.7. *For a Moran construction of a fractal set F satisfying the gap conditions* (3.25) *and* (3.26) *let $f : F \to F$ be such that $f \circ \chi = \chi \circ T$. Let μ be a f-invariant ergodic Borel probability measure with positive entropy, and $\tilde{u} \circ \chi = -\log\lambda$. Then the limit*

$$d_{\mu,q}(x) := \lim_{\varepsilon\to 0^+} \frac{\log \mu(B(x,\varepsilon)) + q\tau(B(x,\varepsilon))}{\log \varepsilon}$$

exists for μ-almost all $x \in F$, and it coincides with

$$\alpha^\nu(q) = \frac{h(\nu) - q}{\int_F \tilde{u}(x)\, d\nu(x)}.$$

This result also holds for transformations satisfying some hyperbolicity condition like in [114], where for instance the following result is proved.

THEOREM 12.8. *Let f be a surface diffeomorphism, and μ and f-invariant ergodic measure with $h(\mu) > 0$ and Lyapunov exponents $\lambda_1 > \lambda_2$. If* supp(μ) *is a compact locally maximal hyperbolic set for f, then*

$$\alpha^\mu(q) = \big(h(\mu) - q\big)\left(\frac{1}{\lambda_1} - \frac{1}{\lambda_2}\right).$$

Chapter 13

The Variational Principle

Let us remind that one says [97,118,119] that the Carathéodory dimension $\alpha_c(Z)$ admits the variational principle (with respect to $\mathcal{M}$) if $\sup_{\mu\in\mathcal{M}} \dim_c \mu = \alpha_c(Z)$, where the supremum is taken over some set $\mathcal{M}$ of measures (it could be Borel probability measures, invariant measures with respect to a dynamical system, etc.). We shall show that indeed the spectrum of dimensions for Poincaré recurrences admits the variational principle.

13.1. Preliminaries and motivation

In the framework of this book a variational principle refers to a relationship between dimension-like quantities, so that a particular dimension coincides with the extremal value of a collection of dimensions of other kind. In this sense, if we had an alternative definition of the spectrum of a measure, then the definition we gave above,

$$\alpha^{\mu}(q) := \inf\{\alpha(q, X') \colon X' \subset X \text{ and } \mu(X') = 1\},$$

could be understood as a variational principle, or more precisely, an inverse variational principle. Instead, in this chapter we deal with a direct variational principle where a "topological dimension" is obtained as the supremum of a set of "measure theoretical dimensions". In order to illustrate this, let us consider the example we developed in the previous section.

Let us remind that for the T-invariant measure μ on X,

$$h_{\mathrm{top}}(\mu) = \inf\{h_{\mathrm{top}}(X') \colon X' \subset X \text{ and } \mu(X') = 1\}.$$

We proved above (Proposition 12.1) that $h_{\mathrm{top}}(\mu) = h(\mu)$, which can be seen as an inverse variational principle. A direct variational principle also holds, indeed,

$$h_{\mathrm{top}}(T) = \sup\{h_{\mathrm{top}}(\mu) \colon \mu \text{ is } T\text{-invariant}\}.$$

This equality is a direct consequence of the variational principle for the topological pressure, as it appears in the thermodynamical formalism [109]. A more general version of the variational principle for the topological pressure is presented [97], and allows one to treat non-compact invariant sets. In our case, since the fractal set resulting of a generalized Moran construction, the hyperbolic repellers and other examples we are considering, are compact, we may be satisfied with the particular version that follows.

THEOREM 13.1. *Let Ω be a specified subshift of the full shift Ω_p. Then for each Hölder continuous function $\psi : \Omega \to \mathbb{R}$,*

$$P_\Omega(\psi) = \sup\left\{h(\mu) + \int_\Omega \psi\, d\mu:\ \mu \text{ is } \sigma\text{-invariant}\right\}.$$

Here $P_\Omega(\psi)$ is the topological pressure associated to the potential ψ, as it was defined by (2.10), or alternatively by (2.17). For the sake of completeness and also for pedagogical reasons, we will give a proof of this result following the classical lines (which can be found for instance in [109]).

PROOF. Let μ be a σ-invariant probability measure. Since ψ is Hölder continuous, there exist θ and $C > 0$ such that

$$\max\left\{\left|\psi(\omega) - \psi(\varpi)\right|:\ \omega, \varpi \in c\right\} \leqslant C\theta^n$$

for each $n \in \mathbb{N}$ and $c \in \zeta^n$. Then, for each $n \in \mathbb{N}$

$$\begin{aligned}\int_\Omega \psi(\omega)\, d\mu(\omega) &= \frac{1}{n}\sum_{j=0}^{n-1}\int_\Omega \psi\left(\sigma^j\omega\right) d\mu(\omega)\\ &= \frac{1}{n}\int_\Omega \psi_n\left(\zeta^n(\omega)\right) d\mu(\omega) \pm \frac{1}{n}\left(\frac{C\theta}{1-\theta}\right)\\ &= \frac{1}{n}\sum_{c\in\zeta^n}\psi_n(c)\mu(c) \pm \frac{1}{n}\left(\frac{C\theta}{1-\theta}\right).\end{aligned}$$

Here we employ the traditional notations $\psi_n(c) := \max_{\omega\in c}\sum_{j=0}^{n-1}\psi(\sigma^j\omega)$, and $A = B \pm C$ for the inequalities $B - C \leqslant A \leqslant B + C$.

Now, for each $n \in \mathbb{N}$ let us define the quantities

$$P_n(\psi, \Omega) := \frac{1}{n}\log\left(\sum_{c\in\zeta^n}\exp\left(\psi_n(c)\right)\right),$$

$$h_n(\mu) := -\frac{1}{n}\sum_{c\in\zeta^n}\mu(c)\log\mu(c),$$

which, as we have already seen, are such that $P(\psi, \Omega) = \lim_{n\to\infty} P_n(\psi, \Omega)$ and $h(\mu) = \lim_{n\to\infty} h_n(\mu)$.

By concavity of the function $x \mapsto \log(x)$, we have that

$$\sum_{c\in\zeta^n} \log\Big(\frac{\exp(\psi_n(c))}{\mu(c)}\Big)\mu(c) \leqslant \log\Big(\sum_{c\in\zeta^n} \exp\big(\psi_n(c)\big)\Big),$$

therefore

$$h_n(\mu) + \int_\Omega \psi(\omega)\, d\mu(\omega) \leqslant P_n(\psi, \Omega) \pm \frac{1}{n}\Big(\frac{C\theta}{1-\theta}\Big),$$

for all $n \in \mathbb{N}$. Taking the limit $n \to \infty$ we obtain

$$h(\mu) + \int_\Omega \psi(\omega)\, d\mu(\omega) \leqslant P(\psi, \Omega),$$

for every σ-invariant Borel probability measure, and in this way we obtain a legal proof of the inequality

$$P(\psi, \Omega) \geqslant \sup\Big\{h(\mu) + \int_\Omega \psi(\omega)\, d\mu(\omega)\colon\ \mu \text{ is } \sigma\text{-invariant}\Big\}.$$

The converse inequality is less evident.

For each $n \in \mathbb{N}$ consider the atomic σ-invariant probability measure

$$\mu_{\psi,n} := \sum_{\omega\in \mathrm{Per}_n(\sigma)} \frac{\exp(\psi_n(\omega))}{\sum_{\varpi\in\mathrm{Per}_n(\sigma)} \exp(\psi_n(\varpi))}\, \delta_\omega,$$

where $\psi_n(\omega) := \sum_{j=0}^{n-1} \psi(\sigma^j\omega)$, $\mathrm{Per}_n(\sigma) := \{\omega \in \Omega\colon \sigma^n\omega = \omega\}$, and δ_ω is the atomic probability measure concentrated at ω. The set of σ-invariant Borel probability measures is convex and compact with respect to the weak topology (see [49] for details). Let

$$\mathcal{E}_\psi := \big\{\mu\colon\ \mu \ \text{ is a limit point of the sequence } \{\mu_{\psi,n}\}_{n=1}^\infty\big\}.$$

Let n_0 be the specification length of X, and for each $n \in \mathbb{N}$ let

$$Z_n := Z_n(\psi, \Omega) = \sum_{c\in\zeta^n} \exp\big(\psi_n(c)\big).$$

For each $n \in \mathbb{N}$, $c \in \zeta^n$ and $m \geqslant n + 2n_0$, the Hölder continuity and the specification property imply that

$$\sum_{\omega\in\mathrm{Per}_m(\sigma)} e^{\psi_m(\omega)} = Z_{m-2n_0}\, Z_n\, e^{\pm 2(\|\psi\| n_0 + \frac{C\theta}{1-\theta})},$$

$$\sum_{\omega\in\mathrm{Per}_m(\sigma)\cap c} e^{\psi_m(\omega)} = Z_{m-2n_0}\, e^{\psi_n(c)} e^{\pm 2(\|\psi\| n_0 + \frac{C\theta}{1-\theta})},$$

where $\|\psi\| := \max_{\omega\in\Omega} |\psi(\omega)|$, and $A = B\,e^{\pm C}$ stands for the inequalities $B\,e^{-C} \leqslant A \leqslant B\,e^{C}$. Then, for each $n \in \mathbb{N}$ and $m \geqslant n + 2n_0$ we have

$$\begin{aligned} h_n(\mu_{\psi,m}) &= -\frac{1}{n}\sum_{c\in\zeta^n} \mu_{\psi,m}(c)\log\left(\frac{\sum_{\omega\in\mathrm{Per}_m(\sigma)\cap c} e^{\psi_m(\omega)}}{\sum_{\varpi\in\mathrm{Per}_m(\sigma)} e^{\psi_m(\varpi)}}\right) \\ &= \frac{\log(Z_n)}{n} - \frac{1}{n}\sum_{c\in\zeta^n} \mu_{\psi,m}(c)\big(\psi_n(c) \pm c_0\big) \end{aligned}$$

with $c_0 := 4(\|\psi\| n_0 + C\theta(1-\theta)^{-1})$. Now, consider a subsequence $m_1 < m_2 < \cdots$ such that $\mu_{\psi,m_k} \to \mu \in \mathcal{E}_\psi$ as $k \to \infty$. Then we have

$$h_n(\mu) + \frac{1}{n}\sum_{c\in\zeta^n} \mu(c)\psi_n(c) = \frac{\log(Z_n)}{n} \pm \frac{c_0}{n},$$

for each $n \in \mathbb{N}$. On the other hand, taking for each $c \in \zeta^n$ an arbitrary point $\omega^* \in c$, we obtain,

$$\begin{aligned} \frac{1}{n}\sum_{c\in\zeta^n} \mu(c)\psi_n(c) &= \frac{1}{n}\sum_{c\in\zeta^n} \mu(c)\sum_{j=0}^{n-1} \psi\big(\sigma^j\omega^*\big) \pm \frac{C\theta}{n(1-\theta)} \\ &= \frac{1}{n}\sum_{j=0}^{n-1}\sum_{c\in\zeta^n} \mu(c)\psi\big(\sigma^j\omega^*\big) \pm \frac{C\theta}{n(1-\theta)} \\ &= \frac{1}{n}\sum_{j=0}^{n-1}\left(\int_\Omega \psi\big(\sigma^j\omega\big)\,d\mu(\omega) \pm C\theta^n\right) \pm \frac{C\theta}{n(1-\theta)}. \end{aligned}$$

Since $\mu \in \mathcal{E}_\psi$ is T-invariant, the inequalities above imply that

$$h_n(\mu) + \int_\Omega \psi\big(\sigma^j\omega\big)\,d\mu(\omega) = \frac{\log(Z_n)}{n} \pm \left(\frac{2c_0}{n} - C\theta^n\right),$$

for all $n \in \mathbb{N}$, and taking the limit $n \to \infty$ we finally obtain

$$h(\mu) + \int_\Omega \psi(\omega)\,d\mu(\omega) = P(\psi, \Omega),$$

for every $\mu \in \mathcal{E}_\psi$. This concludes the proof. □

It is important to remark that in the particular case under consideration, the set $\mathcal{E}_\psi$ is composed by a single ergodic measure μ_ψ with positive entropy. This non-trivial result can be found, for instance, in [107].

13.2. A variational principle for the spectrum

Theorem 12.2 establishes a relationship between the measure theoretical entropy of a measure, and the spectrum of dimensions for Poincaré recurrences for the same measure. On the other hand, by Theorem 5.1 the spectrum of dimensions for Poincaré recurrences is the solution of an equation written in terms of the topological pressure of certain potential. In this subsection we will use the variational principle for the topological pressure that we have just presented, to deduce the analogous principle for the spectrum of dimensions for Poincaré recurrences.

THEOREM 13.2. *For a generalized Moran construction of a fractal set F satisfying the gap conditions* (3.25) *and* (3.26) *let $f : F \to F$ be such that $f \circ \chi = \chi \circ \sigma$, and $\tilde{u} \circ \chi = -\log \lambda$. In the region $0 \leqslant q \leqslant h_{\mathrm{top}}(f)$ we have*

$$\alpha_c(q) = \sup\bigl\{\alpha^{\nu}(q) \colon \ \nu \text{ is } f\text{-ergodic with } h(\nu) > 0\bigr\}.$$

PROOF. First, Theorem 5.1 states that $\alpha_c(q)$ is the unique solution of the equation $P(\alpha \log \lambda) = q$, which – thanks to the gap conditions and the topological conjugacy $\chi : \Omega \to F$ – corresponds to the equation $P(-\alpha\tilde{u}) = q$. Therefore, for $q \in [0, h_{\mathrm{top}}(f)]$, $\alpha_c(q)$ is the unique solution of the equation $P(-\alpha\tilde{u}) = q$.

On the other hand, Theorem 12.2 ensures that for each f-ergodic Borel probability measure ν in F such that $h(\nu) > 0$, the quantity $\alpha^{\nu}(q)$ is the unique solution of the equation

$$q = h(\nu) - \int_F \alpha\tilde{u}(x)\, d\nu(x).$$

The variational principle for the topological pressure tell us that

$$h(\nu) - \int_F \alpha\tilde{u}(x)\, d\nu(x) \leqslant P(-\alpha\tilde{u}),$$

for every f-invariant Borel probability measure, and, in particular, for each f-ergodic Borel probability measure with positive entropy. Thus

$$P\bigl(-\alpha_c(q)\tilde{u}\bigr) = q \geqslant h(\nu) - \int_F \alpha_c(q)\tilde{u}(x)\, d\nu(x),$$

$$q = h(\nu) - \int_F \alpha^{\mu}(q)\tilde{u}(x)\, d\nu(x),$$

therefore

$$\int_F \bigl(\alpha_c(q) - \alpha^{\nu}(q)\bigr)\tilde{u}(u)\, d\nu(x) \geqslant 0.$$

Since $\tilde{u}(x) > 0$ for each $x \in F$, we obtain the inequality

$$\alpha_c(q) \geqslant \sup\{\alpha^{\nu}(q)\colon \ \nu \text{ is } f\text{-ergodic with } h(\nu) > 0\}.$$

Associated to the Hölder continuous function $\omega \mapsto \alpha_c(q)\log\lambda(\omega)$ in Ω, define $\mathcal{E}_{\alpha_c(q)\log\lambda}$ as we did in the proof of the variational principle. As we mentioned above, a result from Ruelle's (Theorem 2.1 in [107]) implies that $\mathcal{E}_{\alpha_c(q)\log\lambda} := \{\mu_{\alpha_c(q)\log\lambda}\}$, and $\mu := \mu_{\alpha_c(q)\log\lambda}$ is σ-ergodic with positive entropy. The corresponding measure $\nu := \mu \circ \chi^{-1}$ on F satisfies

$$h(\nu) - \int_F \alpha_c(q)\tilde{u}(x)\,d\nu(x) = P(-\alpha_c\tilde{u}).$$

Since χ is a measure theoretical isomorphism, the induced measure $\nu = \mu \circ \chi^{-1}$ inherits ergodicity and has positive entropy as well. With this we conclude the proof. □

13.3. The variational principle for suspended flows

In Chapter 10 we computed the spectrum for Poincaré recurrences for a suspended flow (X^{ϕ}, Φ) over a specified subshift (Ω, σ). Let us rephrase the main result of that chapter:

For $\alpha \geqslant 1$ and $q \geqslant 0$, the spectrum for Poincaré recurrences $\alpha^* \equiv \alpha(X^{\phi}, q)$ satisfies the equation

$$P_X\big((1-\alpha^*)u - q\phi|\sigma\big) = 0.$$

With this and Theorem 12.3 we just proved, we obtain the following

THEOREM 13.3. *Let $\mathcal{M}_e$ be the set of all ergodic Φ-invariant probability measures in X^{ϕ}. Then, if $\alpha(X^{\phi}, q) \geqslant 1$, $q \geqslant 0$, we have*

$$\alpha\big(X^{\phi}, q\big) = \sup\{\alpha^{\bar{\mu}}(q)\colon \ \bar{\mu} \in \mathcal{M}_e,\ h(\mu) > 0\},$$

where μ is the σ-invariant probability measure projected on Ω from $\bar{\mu}$.

The proof of this result follows the same lines as the proof of Theorem 13.2 above, substituting α by $\alpha - 1$.

PART V

PHYSICAL INTERPRETATION AND APPLICATIONS

Chapter 14

Intuitive Explanation of Some Notions and Results of this Book

In spite of sophisticated calculations and estimates in the proofs of theorems of our book, the main ideas are simple and natural for people who are familiar enough with statistical mechanics machinery. The goal of this chapter is to explain on a physical level of rigorousness (in the spirit of the articles [43,47,66]), why and how topological pressure, topological entropy, Lyapunov exponents and spectra of dimensions for Poincaré recurrences fuse together to form Bowen-type equations.

14.1. Topological entropy, Lyapunov exponents and Poincaré recurrences for ergodic conformal repellers

The Hausdorff dimension reflects only geometrical features of a chaotic motion. It does not expose characteristic dynamical behavior. Nevertheless, we know that, for some systems, Hausdorff dimensions are determined by dynamical quantities. This fact is illustrated in the following examples.

14.1.1. Entropy

Consider the simple situation when (X, f, μ) is an ergodic conformal repeller with $X \subset \mathbb{R}^m$ and μ a f-invariant, normalized and ergodic measure. Let us remind that a smooth map $f : X \to X$ is conformal if at every point $x \in X$ we have that $Df(x) = L(x)\mathrm{Isom}_x$, where $L(x)$ is a number and Isom_x is an isometry. The map f is expanding at x if $|L(x)| > 1$ and is contracting at x if $|L(x)| < 1$. A repeller for such a map f (i.e., an invariant locally maximal repelling set) is called a conformal repeller.

Let $\mathcal{G} = \{G_1, \dots, G_p\}$ be a cover of X of $\operatorname{diam}\mathcal{G} = \varepsilon$, where $\operatorname{diam}\mathcal{G} := \max\{\operatorname{diam} G_1, \dots, \operatorname{diam} G_p\}$. The n-itinerary of point x relative to $\mathcal{G}$ is the word

$$\omega_x(n) := i_0 i_1 \dots i_j \dots i_{n-1}, \quad i_j \in \{1, \dots, p\},$$

such that $f^j(x) \in G_{i_j}$ for each $j = 0, \dots, n-1$. A cylinder set of the length n is the collection of all points having the same n-itinerary:

$$\zeta_n(x) := \{y \in X\colon \omega_x(n) = \omega_y(n)\}.$$

Assume that the collection of all cylinder sets is a generating cover such that for longer lengths n, $\mathcal{G}^n := \{\zeta_n(x)\colon x \in X\}$ provides a finer cover of X with $\operatorname{diam} \mathcal{G}^n =: r(n, \varepsilon) \to 0$ as $n \to \infty$. By Shannon's theorem we know that there exists a number $h_\mu(f) \geqslant 0$ such that $\#\mathcal{G}^n \sim e^{nh_\mu(f)}$ as $n \to \infty$.[2] Thus, μ is the measure of maximal entropy and we have that $h_\mu(f) = h_{\text{top}}(f) =: h(f)$. Furthermore, by assumption, the full measure $\mu(X) = 1$ is shared equitably among cylinder sets such that entropy yields the following estimate for the measure of a cylinder set of the length n

$$\mu(\zeta_n(x)) = \frac{1}{\#\mathcal{G}^n} \sim e^{-nh(f)}.$$

14.1.2. Lyapunov exponents

The diameter $r(n, \varepsilon)$ of the cover $\mathcal{G}^n$, as a function of n, can be estimated by the average expansiveness of the map f along the orbit. For a point $x \in X$ let

$$\lambda_n(x) = \frac{1}{n} \sum_{j=0}^{n-1} \log\bigl|L(f^j(x))\bigr|. \tag{14.1}$$

The limit $\lambda(x) = \lim_{n\to\infty} \lambda_n(x)$ exists for x a μ-typical point. Because of the ergodicity assumption the limit above exists and is a constant, λ, independent on the choice of typical point x. Assuming the uniform contraction rate $\exp(-\lambda)$ for cylinders, the diameter of the collection $\mathcal{G}^n$ of cylinder sets is estimated to be

$$r(n, \varepsilon) \approx \varepsilon \prod_{j=0}^{n-1} \bigl|L(f^j(x))\bigr|^{-1} \approx \varepsilon e^{-n\lambda}. \tag{14.2}$$

If, for a sufficiently long n, the orbits are distributed uniformly in X, the measure of cylinders may be estimated by the measure of a ball of radius $r = r(n, \varepsilon)$ as follows

$$\mu\{y \in X\colon d(f^k x, f^k y) < \varepsilon,\ k = 0, \dots, n-1\} \sim \mu(B(x, r)).$$

Thus,

$$\mu(B(x, r)) \sim e^{-nh(f)} \sim \left(\frac{r}{\varepsilon}\right)^{h(f)/\lambda} \quad \text{as } r \to 0$$

[2] By $f(n) \sim g(n)$ we mean here that there exist positive constants $\underline{C}$ and $\overline{C}$ such that $\underline{C}g(n) \leqslant f(n) \leqslant \overline{C}g(n)$ for sufficiently large n.

which tells us that

$$\dim \mu = \frac{h(f)}{\lambda} = \inf\{\dim_H A\colon \mu(A) = 1\} = \dim_H X, \tag{14.3}$$

since for the ergodic conformal repeller the local (pointwise) and the global dimensions of μ coincide. Thus, a geometric property of the ergodic measure μ is determined entirely by two dynamical quantities.

14.1.3. *The spectrum of dimensions for Poincaré recurrences*

Let us follow lines similar to those used in the computation of the Hausdorff dimension. So, let us distinguish all the subsets B_i, $i = 1, 2, \ldots, N(n)$, of the invariant subset X having the Poincaré recurrence equals to n, i.e.,

$$\tau(B_i) := \min\{t \in \mathbb{N},\ x \in B_i\colon\ f^n(x) \in B_i\} = n \quad \text{for } i = 1, 2, \ldots, N(n). \tag{14.4}$$

Recall that we are considering an ergodic conformal repeller. Thus, we expect that $N(n)$, the number of sets having Poincaré recurrence equals to n, grows exponentially fast, while the diameters shrink exponentially fast to zero (see (14.2)). Otherwise said, there should be constants $h_0 > 0$, and $0 < \lambda_{\min} \leqslant \lambda_{\max} < 1$, for which

$$N(n) \approx e^{nh_0} \quad \text{and} \quad \lambda_{\min}^n \leqslant \operatorname{diam}(B_i) \leqslant \lambda_{\max}^n.$$

The exponentials have opposite behaviors. Thus, depending on the value of number α, the sum $\sum_{i=1}^{N(n)}(\operatorname{diam}(B_i))^\alpha$ may either diverge to infinity or converge to zero exponentially fast. There are cases which we study in detail, where collections $\{B_i,\ i = 1, 2, \ldots, N(n)\}$ of sets with Poincaré recurrence equals to n essentially cover the attractor, so that

$$\sum_{i=1}^{N(n)}\bigl(\operatorname{diam}(B_i)\bigr)^\alpha \sim 1 \quad \text{for } \alpha = \dim_H(X).$$

Let us introduce a parameter q in the role of a time scale to gauge the "importance" of different Poincaré recurrences. "The importance" will be a numeric characteristic of the set X, denoted $\alpha_c(q)$, of the same kind as its Hausdorff dimension, that is such that

$$\sum_{i=1}^{N(n)} \operatorname{diam}(B_i)^{\alpha(q)} e^{-q\tau(B_i)} \approx 1. \tag{14.5}$$

The spectrum of dimensions for Poincaré recurrences is the functional dependence $\alpha(q)$ of the dimension-like characteristic α with respect to the time scale q.

This spectrum can be seen as a relation between scale transformations in time and scale transformations in space. These changes are to be made in the logarithmic scale for the space, while they could be linear or logarithmic for the time, depending on the recurrence properties of the system.

For an ergodic conformal repeller X the diameter of sets B_i in (14.5) is estimated in (14.2) and each set has Poincaré recurrence n. Thus, the spectrum of dimensions for Poincaré recurrences is determined by the following relation,

$$\sum_{B_i} e^{n(-q+\alpha(q)\lambda_n(x_i))} \sim 1, \quad n \gg 1, \tag{14.6}$$

where $x_i \in B_i$. The number of sets B_i, with Poincaré recurrence n, may be estimated by the number of periodic points of the period n. For the repeller X of the conformal map f we take

$$\#\mathrm{Per}_n(f) \approx e^{nh},$$

where $\mathrm{Per}_n(f)$ is the set of all periodic points of minimal period n and h is the topological entropy. Thus, relation (14.6) becomes

$$e^{n(h-q+\alpha(q)\lambda)} \sim 1, \quad n \gg 1,$$

which holds for

$$\alpha(q) = \frac{h-q}{\lambda}. \tag{14.7}$$

Recall that, by assumption, the system (X, f) is ergodic and then $\lambda_n(x) \approx \lambda$, $n \gg 1$, independent of a typical point x. The spectrum for Poincaré recurrences is then related to the Hausdorff dimension of the ergodic conformal repeller X by substituting relation (14.3) into (14.7): we obtain

$$\alpha(q) = \dim_H(X)\left(1 - \frac{q}{h}\right). \tag{14.8}$$

14.2. (Non-ergodic) Conformal repellers

The results in the previous section are not valid in general. In relation (14.3), for instance, the limit $\lambda := \lim \lambda_n(x)$ is not independent of point $x \in X$ when the repeller is not ergodic. If we want to follow as before the same arguments in more general situations then we must perform a multifractal analysis by decomposing the set X into level subsets, collecting all points with the same value for $\lim \lambda_n(x)$. We shall see that for such level sets a relation of the type (14.3) is still valid for non-ergodic conformal repellers. However, there are no simple expressions, as (14.7) and (14.8), for the spectrum of Poincaré recurrences. Instead, we get Bowen-type equations.

14.2.1. *The entropy spectrum for Lyapunov exponents*

Generally, when $\lim_{n\to\infty} \lambda_n(x)$ exists it takes value in the closed interval $[\lambda_1, \lambda_2]$, for some $\lambda_1 \leqslant \lambda_2$. In the previous example, Section 14.1, the situation was that $\lambda_1 = \lambda_2 =: \lambda$. Thus, to follow a line of reasoning similar to the one we followed for ergodic conformal repellers, let us first decompose X into level sets of constant Lyapunov exponent,

$$X_\lambda = \{x\colon \lim \lambda_n(x) = \lambda\}, \tag{14.9}$$

when the limit exists. The collection $\mathcal{G}^n$ of cylinder sets has to be trimmed too, according to the range of values of $\lambda_n(x)$, $n \gg 1$. So, let $J \subset [\lambda_1, \lambda_2]$ be an open interval and let

$$\mathcal{G}^n(J) = \{\zeta \in \mathcal{G}^n\colon \exists x \in \zeta \text{ with } \lambda_n(x) \in J\}.$$

Then, for the given interval $J \subset [\lambda_1, \lambda_2]$, a "partial entropy" function would be

$$\eta(J) = \lim_{n\to\infty} \frac{1}{n} \log \#\mathcal{G}^n(J).$$

The *entropy spectrum for Lyapunov exponents* is then defined to be

$$\eta(\lambda) \equiv h_{\text{top}}(f|X_\lambda) = \inf\{\eta(J)\colon J \ni \lambda\}. \tag{14.10}$$

Then, waving hands in a similar way as we did in Section 14.1 we find that

$$\dim_H X_\lambda = \frac{\eta(\lambda)}{\lambda}.$$

14.2.2. *The spectrum of dimensions for Poincaré recurrences*

Relation (14.2) is not valid for the present situation. Then, in order to estimate the diameters of the sets $\{B_i,\ i = 1, 2, \ldots, N(n)\}$ having a Poincaré recurrence equal to n, we associate to each of those sets a periodic point of minimal period n. In this way every B_i would be a neighborhood of a particular periodic point x_i in the repeller $X \subset \mathbb{R}^m$, such that $f^n x_i = x_i$ and $f^k x_i \neq x_i$ for $k < n$. Suppose that the n points in the orbit of x_i are well distributed in the attractor, so that

$$\min\{\operatorname{dist}(f^k x_i, f^\ell x_i)\colon 0 \leqslant k \leqslant \ell < n\} \approx \frac{1}{n}\operatorname{diam}(X).$$

In this case, in order to ensure that the image sets $f^k B_i$, $k = 1, 2, \ldots, n-1$, of a neighborhood of x_i, do not contain x_i, it is enough that

$$\begin{aligned}\operatorname{diam}(f^k B_i) &\leqslant \operatorname{dist}(f^k x_i, x^i)\\ &\leqslant \min\{\operatorname{dist}(f^k x_i, f^\ell x_i)\colon 0 \leqslant k \leqslant \ell < n\} \approx \frac{\operatorname{diam}(X)}{n}.\end{aligned}$$

An upper bound for the diameters of the sets $f^k B_i$ is obtained by using the maximal Lyapunov exponent through the orbit of x_i. Indeed, by taking

$$\operatorname{diam}(B_i) \approx \frac{\operatorname{diam}(X)}{n} \exp\left(-\sum_{k=0}^{n-1} \log\left|L\left(f^k x_i\right)\right| \right), \tag{14.11}$$

where $\log|L(x)|$ is the Lyapunov exponent at x of the conformal map f, which is supposed to be positive. We have that

$$\operatorname{diam}\left(f^k B_i\right) \approx \frac{\operatorname{diam}(X)}{n} \exp\left(-\sum_{\ell=k}^{n-1} \log\left|L\left(f^\ell x_i\right)\right| \right) \leqslant \frac{\operatorname{diam}(X)}{n},$$

for each $0 \leqslant k < n$. Thus, from relation (14.5) defining the spectrum of dimensions for Poincaré recurrences, taking into account (14.11), we obtain the asymptotic relation

$$\sum_{x \in \operatorname{Per}_n(f)} \exp\left(-\sum_{k=0}^{n-1} \left(q + \alpha(q) \log\left|L\left(f^k x\right)\right|\right) \right) \sim 1, \tag{14.12}$$

where $\operatorname{Per}_n(f)$ is the set of all periodic points of minimal period n. We can alternatively write this relation as

$$\frac{1}{n} \log \sum_{x \in \operatorname{Per}_n} \exp\left(-\alpha(q) \sum_{k=0}^{n-1} \log\left|L\left(f^k x\right)\right| \right) \approx q.$$

In the case of a strongly chaotic system (compare to the result about the topological entropy for expansive transformations in [49], p. 110), the quantity at the left hand side of the previous relation converges as n goes to infinity to a dimension-like characteristic, denoted by $P(-\alpha(q) \log|L|)$, known as the topological pressure for the potential

$$\varphi(x) := -\alpha(q) \log\left|L(x)\right|$$

(see Section 2.4 for a more detailed explanation). Hence, we have the formula

$$P\left(-\alpha(q) \log|L|\right) = q \tag{14.13}$$

that is satisfied by the spectrum of dimensions for Poincaré recurrences.

Considered as a function of the parameter α, the topological pressure $P(-\alpha \log|L|)$ is convex and decreasing. For $\alpha = 0$ it coincides with the topological entropy of the system, which in our case is given by the limit

$$h_{\text{top}} = \lim_{n \to \infty} \frac{1}{n} \#\operatorname{Per}_n(f).$$

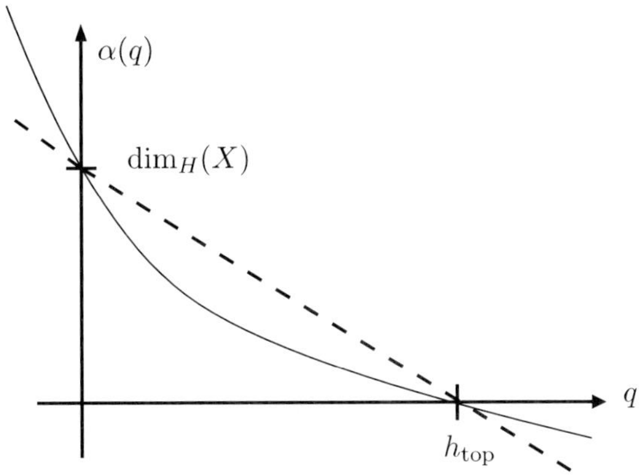

Figure 14.1. The convex shape of the spectrum of dimensions for Poincaré recurrences. The dashed line is the spectrum (14.8) for the ergodic conformal repeller.

For $\alpha = \dim_H(X)$, we necessarily have $P(-\alpha L) = 0$ since in that case the sum in (14.5) is the same as one uses in the computation of the Hausdorff dimension, in which the recurrence times do not appear. Thus, the spectrum $\alpha(q)$ is a convex decreasing function of q, such that $\alpha(0) = \dim_H(X)$ and $\alpha(h_{\text{top}}) = 0$. The spectrum of dimensions for Poincaré recurrences is sketched in Figure 14.1.

14.2.3. A Legendre-transform pair

In the estimates we performed in the previous section, which are valid in the case of a strongly chaotic system, we dealt with the sum

$$\Sigma_n := \sum_{x \in \operatorname{Per}_n(f)} \exp\left(-\alpha(q) \sum_{k=0}^{n-1} \log\left| L\left(f^k x\right)\right| \right).$$

For each periodic point $x \in \operatorname{Per}_n(f)$, we consider the orbit average (14.1) which take values in the interval $[\lambda_1, \lambda_2]$. Let $\{\lambda_1 = \ell_0 < \ell_1 < \cdots < \ell_n = \lambda_2\}$ be an equally spaced partition of the interval $[\lambda_1, \lambda_2]$. Let $J_k = [\ell_{k-1}, \ell_k]$, for $k = 1, \ldots, n$. To each of the intervals J_k we associate the collection of periodic points

$$E_n(J_k) := \left\{ x \in \operatorname{Per}_n(f) \colon \lambda_n(x) \in J_k \right\}. \tag{14.14}$$

From this, we have

$$\Sigma_n \approx \sum_{k=0}^{n-1} \exp\left(n \left(-\alpha(q)\ell_k + \frac{1}{n} \log \#E_n(J_k) \right) \right)$$

$$= e^{nq_n^*(\alpha)} \sum_{k=0}^{n-1} \exp\left(n\left(-\alpha(q)\ell_k + \frac{1}{n}\log \#E_n(J_k) - q_n^*(\alpha)\right)\right),$$

where

$$q_n^*(\alpha) := \max_k \left(-\alpha(q)\ell_k + \frac{1}{n}\log \#E_n(J_k)\right). \tag{14.15}$$

Since

$$1 \leqslant \sum_{k=0}^{n-1} \exp\left(n\left(-\alpha(q)\ell_k + \frac{1}{n}\log \#E_n(J_k) - q_n^*(\alpha)\right)\right) \leqslant n,$$

then (14.12) can be written as

$$A(n)e^{(q_n^*(\alpha)-q)n} \sim 1,$$

with $1 \leqslant A(n) \leqslant n$ and $q^*(\alpha)$ as above. Therefore, in the limit $n \to \infty$ we necessarily have

$$q = \lim_{n\to\infty} q^*(\alpha) \quad \text{for } \alpha = \alpha(q). \tag{14.16}$$

This condition defines the spectrum of dimensions for Poincaré recurrences. To write it in the familiar form, notice that, for a strongly chaotic system, the function

$$\eta(\lambda) := \lim_{n\to\infty} \frac{1}{n}\log \#E_n(J_k) \quad \text{such that } J_k \ni \lambda,$$

if it is well defined on the interval $[\lambda_1, \lambda_2]$, coincides with the entropy spectrum for Lyapunov exponents, defined in (14.10). Thus, condition (14.16) can be written in the form

$$\begin{aligned} q &= \lim_{n\to\infty} \max_k \left(-\alpha(q)\ell_k + \frac{1}{n}\log \#E_n(J_k)\right) \\ &= \max_\lambda \left(-\alpha(q)\lambda + \eta(\lambda)\right). \end{aligned} \tag{14.17}$$

Notice that the right hand side of this equation is the Legendre transform of the function $-\eta(\lambda)$ evaluated at $-\alpha(q)$ (see, for instance, Chapter VI of [51]), i.e.,

$$q(\alpha) = (-\eta)^*(-\alpha), \tag{14.18}$$

where, as usual, g^* denotes the Legendre transform of the function g. Since the Legendre transform is an involution, we have that

$$-\eta(\lambda) = q^*(-\alpha) = \max_\alpha(\alpha\lambda - q), \quad \text{where } \alpha = \alpha(q),$$

and from this it follows that

$$\eta(\lambda) = \min_\alpha(\alpha\lambda + q) \quad \text{for } q = q(\alpha). \tag{14.19}$$

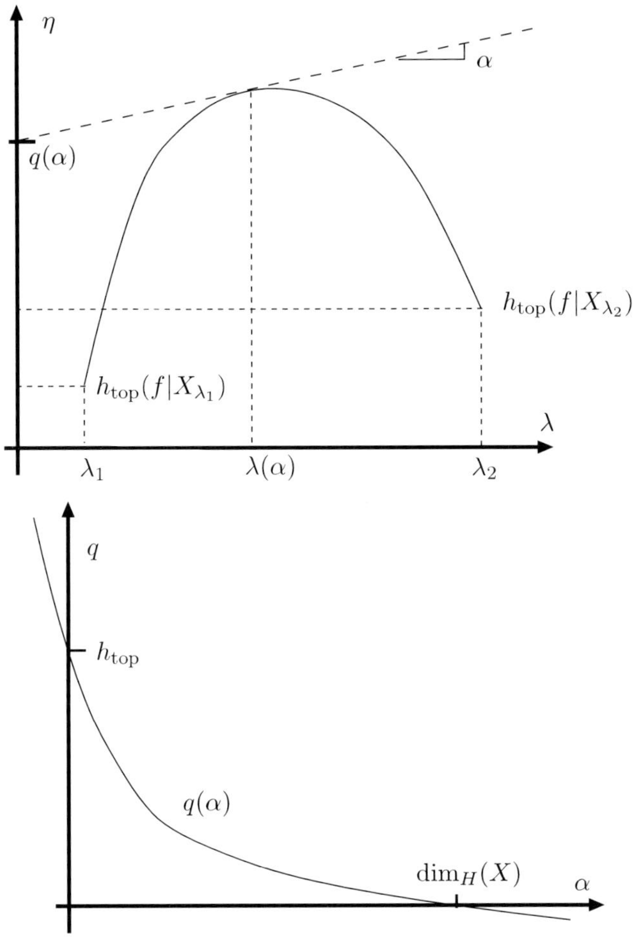

Figure 14.2. The concave shape of the entropy spectrum of Lyapunov exponents, and the inverse $q = q(\alpha)$ of the spectrum of dimensions for Poincaré recurrences.

The relation (14.18) between the inverse of the spectrum of dimensions for Poincaré recurrences, $q(\alpha)$, and the entropy spectrum for Lyapunov exponents, $\eta(\lambda)$, is sketched in Figure 14.2.

Thus, in such a nice situation, as a conformal ergodic repeller, the spectrum of dimensions for Poincaré recurrences and the entropy spectrum of Lyapunov exponents are not independent – in fact, they contain similar information about chaotic features of the system.

Chapter 15

Poincaré Recurrences in Hamiltonian Systems

15.1. Introduction

This chapter shows, in an example, that the dimension for Poincaré recurrences is an important characteristic for Hamiltonian systems with dynamical behaviors that are (multi)fractal in space and time, simultaneously. We will follow mainly the work [12].

There are many fractal objects in the phase space of a Hamiltonian system that reflect complexity of behaviour of its orbits, see [41,50,72,81,84,85,88,127]. Let us mention only such objects as cantori and islands-around-islands (see review [83,84]). Concerning dynamics, the motion is not ergodic in the full phase space, and one needs to restrict dynamics to one of those invariant subsets. In a system with a structure of islands-around-islands the natural invariant subset is the sticky set. Our understanding of such a structure is summarized in Section 3.5.

We apply a popular notion of multifractal analysis [66,54,47,92] to Poincaré recurrences. An invariant subset (e.g., the sticky set) is decomposed into level sets where Poincaré recurrences scale with the same local exponent. Such a decomposition is related to the spectrum of dimensions for Poincaré recurrences. A main result is a formula that, due to the coupling of space and time performed by dynamics, expresses the dimension for Poincaré recurrences in terms of the fractal dimension of the invariant subset, that is a space characteristic, and the exponent in the asymptotic distribution law for Poincaré recurrences, that reflects the behavior of a system in time.

G.M. Zaslavsky was the first who has discovered and studied this remarkable feature of Hamiltonian dynamics, see for instance [127,128] and references therein.

15.2. Asymptotic distributions of Poincaré recurrences

Distribution function for trajectories in phase space can be fairly uniform, as happens, for example, in Anosov systems. Nevertheless, a typical Hamiltonian system has a rich set of islands in phase space, with a regular dynamics inside the islands

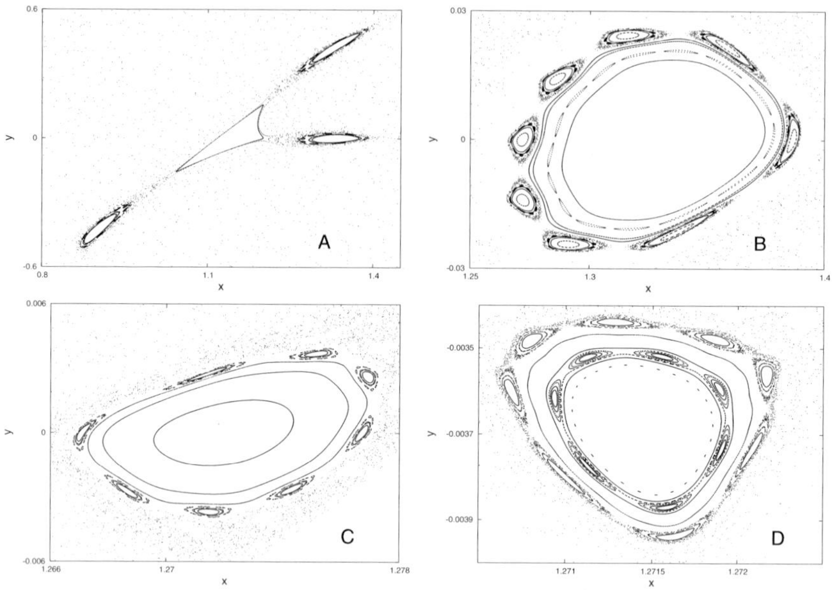

Figure 15.1. Islands around islands in the phase space of the standard map.

and with narrow stochastic layers isolated from the main stochastic sea domain. As an example, the standard map $(x, y) \mapsto s(x, y) = (\bar{x}, \bar{y})$, with

$$\bar{y} = y - K \sin x, \quad \bar{x} = x + \bar{y}, \tag{15.1}$$

has a fairly well known island structure that manifests such kind of behavior. See Figure 15.1, that was obtained in [12].

In Figure 15.1A, we see a phase portrait of the standard map with a structure of four islands: the central one- and three-island resonance sets around the central island. The three-island set separate from the central island as a result of a bifurcation when the parameter K exceeded some critical value. If we continue to increase K, a similar structure of subislands occurs for the three satellite islands of the first generation. In Figure 15.1B, we show a magnification of the right island of the first generation. There is an eight-island chain around it. When the value

$$K = 6.908\,745; \tag{15.2}$$

is reached, it seems that an infinite hierarchy of islands-around-islands exists, with a constant proliferation number of islands $g = 8$ between generations. Figures 15.1C and 15.1D show the two generations following the first generation in the eight-island chain. Our understanding about the infinite hierarchy of islands-around-islands is summarized in Section 3.5.

The dynamics near the islands boundary is singular due to the phenomena of stickiness, and it dominates in the long-time asymptotics. This circumstance influences almost all important probability distributions such as the distribution of distances, exit times, recurrences, moments, etc. The main feature of all such distributions is that they do not correspond to either Gaussian or Poissonian (or similar) processes with all finite moments. This rather manifests in the presence of powerlike tails in the asymptotic limit of distributions for long-time and small-space scales, see [127] and [128].

We are interested in the distribution of Poincaré recurrences. Consider an open ball $B = B(x, r)$ of radius r centered at point x located in the sticky set of an infinite hierarchy of islands around islands. The system is Hamiltonian and the dynamics is area preserving, i.e., Lebesgue's measure μ is invariant, however it is not ergodic. For the trajectory $\{y_i = f^i(y)\colon i \geqslant 0\}$ let $\{i_j\}_{j\geqslant 0}$ be all of the time instants when the trajectory lands in the ball: $y_{i_j} \in B$. The intervals

$$t_j = \{i_{j+1} - i_j\}, \quad j = 0, 1, \ldots, \tag{15.3}$$

are *Poincaré cycles*.

Let us assume that $\mu(X) < \infty$ and let x be located in the sticky set. Denote by W_∞ the collection of all points in $X \setminus B$ that never reach the ball B (see Chapter 17). Then, according to Kac's theorem 17.2, the mean of Poincaré cycles (15.3) for the ball B

$$\bar{\tau}(B) = \frac{\mu(X \setminus W_\infty)}{\mu(B)} < \infty, \tag{15.4}$$

is finite. This allows us to introduce the distribution of Poincaré cycles. For a fixed $\tau > 0$ let

$$P\big(n > \tau \cdot \bar{\tau}(B)\big) = \sum_{k=\lceil \tau \cdot \bar{\tau}(B)\rceil}^{\infty} \frac{\mu(A_k)}{\mu(B)} \tag{15.5}$$

denote the probability to return to the ball B no sooner than in $\tau \cdot \bar{\tau}(B)$ time steps. According to Chapter 17, in definition (15.5) the set $A_k \subset B$ denotes the collection of points that return to the ball, for the first time, in k time steps. Next, we consider the limit $r \to 0$ for a fixed $\tau > 0$.

The chaotic dynamics is considered to be the normal one if the asymptotic law

$$P(\tau) := \lim_{r\to 0} P\big(n > \tau\, \bar{\tau}(B)\big) \sim \exp(-\tau), \tag{15.6}$$

exists and is Poissonian, with all moments finite. The Poissonian limit law has been proved to exist for automorphisms of the torus and Markov chains by Pitskel in [103]. In [67] Hirata proved it for Axiom-A diffeomorphisms and shifts of finite type with a Hölder potential. For piecewise expanding maps of the interval Collet and Galves proved the Poissonian limit law in [44]. However, existence of the limit in (15.6) is not always the case. For homeomorphisms of the circle Coelho and de Faria [42] have found that the asymptotic limit does not exist and distinct asymptotic laws are reached, depending on how the limit $\mu(B) \to 0$ is taken.

The dynamics is said to be an anomalous one when the asymptotic distribution law

$$P(\tau) \sim \tau^{-\gamma} \tag{15.7}$$

is valid.

15.3. A self-similar space-time situation

The sticky riddle $\mathcal{R}$ in the geometric construction of the sticky set in Section 3.5 is illustrated by the Sierpinsky carpet in Figure 15.2. Let the largest square be the only basic set Δ_0 in the zeroth generation. The next generation of basic sets

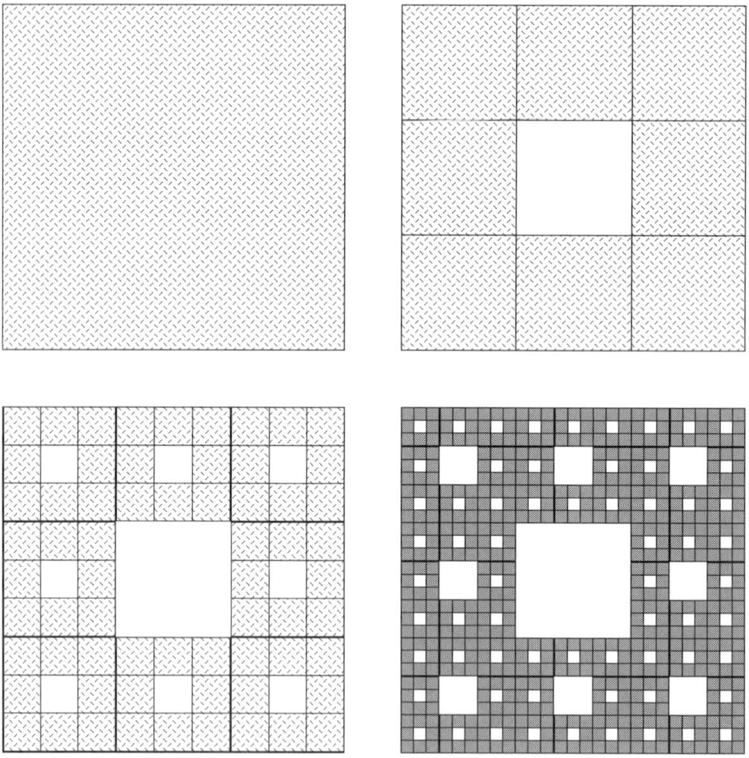

Figure 15.2. Scheme of a self-similar geometric construction of the sticky set of an infinite hierarchy of islands-around-islands structure (see Section 3.5.1).

$\Delta_{\underline{i}_1}$ conform an annulus around the island $\mathcal{I}_0$ (not shown in the figure) which represents its boundary layer. It consists of g_1 ($g_1 = 8$ in Figure 15.2) basic subsets of the first generation, $\Delta_{\underline{i}_1}$. Then, surround each of the first generation basic subsets by an annulus of g_2 ($g_2 = 8$ in Figure15.2) basic sets of the second generation and repeat the process. On the nth step the structure is described by the word $G_n = (g_1, g_2, \ldots, g_n)$. The full number of islands of the nth generation is

$$N_n = g_1 \cdot \cdots \cdot g_n. \tag{15.8}$$

Following Section 3.5, a basic set from the nth generation is labeled by the word

$$\underline{i}_n = (i_1, \ldots, i_j, \ldots, i_n), \quad 1 \leqslant i_j \leqslant g_j, \; j = 1, \ldots, n. \tag{15.9}$$

Then, to each point x of the sticky set corresponds a sequence $\underline{i} = (i_1, i_2, \ldots, i_n, \ldots)$ such that $x \in \Delta_{\underline{i}_n}$ for every prefix $\underline{i}_n$ of the sequence $\underline{i}$.

The time

$$T_{\underline{i}_n} := \min\{t > 0\colon\; \Delta_{\underline{i}_n} \cap s^t(\Delta_{\underline{i}_n}) \neq \emptyset\} \tag{15.10}$$

that a particle takes to return, by the action of the standard map s, to the basic set $\Delta_{\underline{i}_n}$, carries all information of the nth generation of islands. By introducing the return time (15.10) for each basic set, we have attached an additional parameter, describing the temporal behavior, to the simple geometric construction of the sticky set, which is similar to a Cantor set.

A simplified case of the construction described above corresponds to the exact self-similar situation

$$\begin{aligned} S_{\underline{i}_n} &= S_n = \lambda_S^n S_0, \\ T_{\underline{i}_n} &= T_n = \lambda_T^n T_0, \end{aligned} \tag{15.11}$$

for every word $\underline{i}_n$ of a given length n. In conditions (15.11) $S_{\underline{i}_n}$ is the area of the basic set $\Delta_{\underline{i}_n}$ and $T_{\underline{i}_n}$ is the return time introduced in (15.10).

The self-similarity conditions (15.11) correspond to equal areas and equal return times for all basic sets of the same generation. The parameters λ_S and λ_T are the scale factors of the exact self-similarity in space and time, correspondingly, with

$$\lambda_S < 1, \quad \lambda_T > 1. \tag{15.12}$$

In addition to the space-time self-similarity (15.11), we assume self-similarity in the proliferation of islands, i.e.,

$$g_n = g \geqslant 3, \tag{15.13}$$

which corresponds to a situation that is modeled (in the sticky set) by a minimal multipermutative system. It follows from Eqs. (15.8) and (15.13) that

$$N_n = g^n, \quad n \geqslant 0, \tag{15.14}$$

if we start from the only 0th level island ($g_0 = 1$).

Consider now the sequence of boundary layers $\{\Delta_{\underline{i}_n}\}$ as a sequence of coverings of the sticky set. Then consider the statistical sum

$$R_n(q) = \sum_{i_1,\dots,i_n} \left(\frac{1}{T_{\underline{i}_n}}\right)^q = C \sum_{i_1,\dots,i_n} \exp(-nq \ln \lambda_T), \tag{15.15}$$

where C is a positive constant. The number of terms in the sum (15.15) is given in (15.14), and therefore as $n \to \infty$ the covering boundary layer approaches the sticky set and in the limit we obtain that the sum Eq. (15.15) diverges if

$$q < q_0 = \frac{\ln g}{\ln \lambda_T}, \tag{15.16}$$

converges to zero if $q > q_0$, and converges to a positive value if $q = q_0$. The positive number q_0 in (15.16) is the dimension for Poincaré recurrences under the self-similar conditions (15.11). According to Chapter 7, $q_0 = 1$. In our case it is absolutely clear, because of the fact that $\lambda_T = g$, the number of islands.

The number of terms in the sum (15.15) are estimated in terms of the fractal dimension f_{SS} of the sticky set by $\lambda_S^{-nf_{SS}/2}$. Then, an alternative formula for the dimension for Poincaré recurrences is

$$q_0 = \frac{f_{SS}}{2} \frac{|\ln \lambda_S|}{\ln \lambda_T} =: \frac{1}{\beta_0} f_{SS}. \tag{15.17}$$

From (15.16) and (15.17) the fractal dimension of the sticky set results to be

$$f_{SS} = \frac{2 \ln g}{|\ln \lambda_S|}. \tag{15.18}$$

Remark that the fractal dimension f_{SS} involves space parameters, g and λ_S, only; while the dimension for Poincaré recurrences (either (15.16) or (15.17)) involves space as well as time parameters. Parameter β_0 introduced in (15.17) is justified in the next two sections.

15.4. Multifractal analysis by Poincaré recurrences

It was mentioned in Sections 15.1 and 15.2 that Hamiltonian systems with rich sets of islands have a space-time multifractal structure rather than fractal one. Our next purpose is to describe, in the framework of Section 4.3 (see also [66,

54,47,92]), a multifractal spectrum of dimensions that is induced by the Poincaré recurrences.

Consider the partition function

$$R_n(q,\alpha) = \sum_{\underline{i}_n} \left(\frac{1}{T_{\underline{i}_n}}\right)^q \varepsilon_n^\alpha, \tag{15.19}$$

where $\varepsilon_n = \operatorname{diam} \Delta_{\underline{i}_n}$ is independent of the word $\underline{i}_n$.

Let $\underline{i}$ be the code sequence for a point x located in the sticky set. Assume that there is an exponent β such that for each prefix $\underline{i}_n$ of $\underline{i}$

$$T_{\underline{i}_n} \sim \varepsilon_n^{-\beta}, \tag{15.20}$$

as $n \gg \infty$. Then, to deal with a multifractal situation, let $E_n(\beta)$ denote the collection of all basic sets $\Delta_{\underline{i}_n}$ from the nth level for which the scaling law (15.20) holds for the given value of β. In a multifractal situation there are distinct values of exponent β in a closed interval $[\beta_{\min}, \beta_{\max}]$ for which $E_n(\beta) \neq \emptyset$. Moreover $\#E_n(\beta) \sim \varepsilon_n^{-f(\beta)}$, where $f(\beta)$ is the fractal dimension of the level set $E_n(\beta)$. Then, we write

$$dN_n(\beta) = d\beta\, \rho(\beta)(\varepsilon_n)^{-f(\beta)} \tag{15.21}$$

for the number of basic Δ-sets for which the time scaling law is as in (15.20), with exponent between β and $\beta + d\beta$.

Using the multifractal decomposition by return times (15.21) the statistical sum (15.19) is replaced by the integral

$$R_n(q,\alpha) = \int d\beta'\, \rho(\beta')(\varepsilon_n)^{\alpha - f(\beta') + \beta' q}. \tag{15.22}$$

The distribution density $\rho(\beta)$ is a slow function of β. Since we are interested in the limit $\varepsilon_n \to 0$, the integral (15.22) is correctly estimated by the value of β' which makes $q\beta - f(\beta)$ smallest. Thus,

$$R_n(q,\alpha) \sim \varepsilon_n^{\alpha - f(\beta(q)) + q\beta(q)} \tag{15.23}$$

where $\beta(q)$ is the solution to the equation

$$q = f'(\beta). \tag{15.24}$$

Thus, the spectrum of dimensions for Poincaré recurrences, $\alpha(q)$, is given by

$$\alpha(q) = -q\beta(q) + f\big(\beta(q)\big) \tag{15.25}$$

and the dimension for Poincaré recurrences q_0 is a solution to the equation $\alpha(q) = 0$.

If we are in a self-similar situation, as the one introduced in the previous section, equation (15.25) reduces to

$$\alpha(q) = -q\beta_0 + f_{SS}, \tag{15.26}$$

where the exponent β_0 is determined by the scaling law (15.20). The dimension for Poincaré recurrences q_0 was computed already in (15.17). The self-similar model is further analyzed in the next section.

15.5. Critical exponents in the self-similar situation

Assume the self-similar conditions (15.11) hold. Then, there is only one β-level subset in the multifractal decomposition of the sticky set, corresponding to the exponent

$$\beta = \beta_0 := 2\frac{\ln \lambda_T}{|\ln \lambda_S|} \tag{15.27}$$

in the scaling law (15.20). Then the support of the spectral function $f(\beta)$ is the only value of exponent β in (15.27). Comparison of (15.17) with (15.25) tells us that $f(\beta_0)$ coincides with the fractal dimension f_{SS} of the sticky set.

Let us next determine exponent γ in the asymptotic law (15.7). For given integer numbers M and $N > M$, let adapt definition (15.5) to our situation by making $B = \Delta_{\underline{i}_M}$ and $\bigcup_{k\geqslant T_{\underline{i}_N}} A_k = \bigcup_{\underline{i}_N} \Delta_{\underline{i}_N}$. All basic sets $\Delta_{\underline{i}_N}$ at a given level N have equal recurrences, thus

$$P(t \geqslant \tau := T_{\underline{i}_N} S_{\underline{i}_M}) := \frac{g^N S_{\underline{i}_N}}{S_{\underline{i}_M}} \tag{15.28}$$

and by the self-similar condition

$$P(t \geqslant \tau) = g^N {\lambda_S}^{N-M}. \tag{15.29}$$

We shall take the limit $M \to \infty$ by keeping τ constant. This is so if $N = N(M)$ is chosen such that

$$\lambda_S^{-M} = \frac{\lambda_T^N}{\tau} T_0 S_0.$$

Thus,

$$P(t \geqslant \tau) = (g\lambda_T\lambda_S)^N \tau^{-1} T_0 S_0. \tag{15.30}$$

To get the asymptotic behavior of the distribution of recurrences we make $N \to \infty$ by letting τ follow $T_{\underline{i}_N}$, i.e., by letting $N \sim \ln \tau$. Then we get

$$P(\tau) \sim \tau^{-\gamma}, \quad \text{as } \tau \to \infty, \tag{15.31}$$

with exponent

$$\begin{aligned}\gamma &= \frac{|\ln \lambda_S|}{\ln \lambda_T} - \frac{\ln g}{\ln \lambda_T} \\ &= q_0\Big(\frac{2}{f_{SS}} - 1\Big), \end{aligned} \tag{15.32}$$

with $q_0 = 1$ here. The exponent γ of the subexponential tail in the distribution of recurrences is determined by the dimension for Poincaré recurrences and the fractal dimension of the invariant (sticky) subset.

15.6. Final remarks

One feels urged to confront the formula (15.32) with experimental data from an actual Hamiltonian system. Such a comparison would be significative if we are able to prove that the experimental situation (e.g., the one depicted in Figure 15.1) corresponds to a self-similar structure in space and in time, as it was assumed in the discussion that lead us to formula (15.32). Thus, a first problem is to apply the numerical methods of multifractal analysis (e.g. to the standard map system) in order to estimate the β-spectrum, $[\beta_{\min}, \beta_{\max}]$, and the corresponding spectral function of dimensions $f(\beta)$ for multiple values of the system parameters (K for the standard map (15.1)), looking for the system with the narrowest spectrum.

However, by a direct numerical estimate of the scale transformation parameters $(\lambda_S, \lambda_T, g)$ along the island set, such as in [22,126], the existence of a simple fractal situation is argued to exists for the standard map with parameter K in (15.2). For this system the Poincaré cycle distribution was estimated in [12] by collecting recurrence times from 5.5×10^5 initial conditions, each run consisting of 10^6 iterations. The histogram is plotted in Figure 15.3.

The histogram follows a Poissonian law for recurrence times smaller than 1.5×10^4 time steps, and then a crossover to a long powerlike tail takes place. The analysis of the tail gives the exponent $\gamma = 3.2 \pm 0.2$. This result, in combination with formula (15.32), shows that $f_{SS} \approx 0.5$. However, to get an experimental confirmation of formula (15.32) we need an independent numerical estimate of the fractal dimension f_{SS} of the sticky set.

At the moment, we are not familiar with a possibility to extend formula (15.32) to more complicated scenarios such as a multifractal space–time structure of islands-around-islands or chaotic dynamics of dissipative systems, not to mention more sophisticated physical problems such as turbulent flow. Even in the very special structure of islands-around-islands the proliferation of islands can follow a very complicated and non-universal scheme as to allow us reduce the connection between $T_{\underline{i}_n}$ and $S_{\underline{i}_n}$ to a single scaling parameter. Our simplifying assumption is justified only as an approximation in the case of a narrow multifractal β-spectrum.

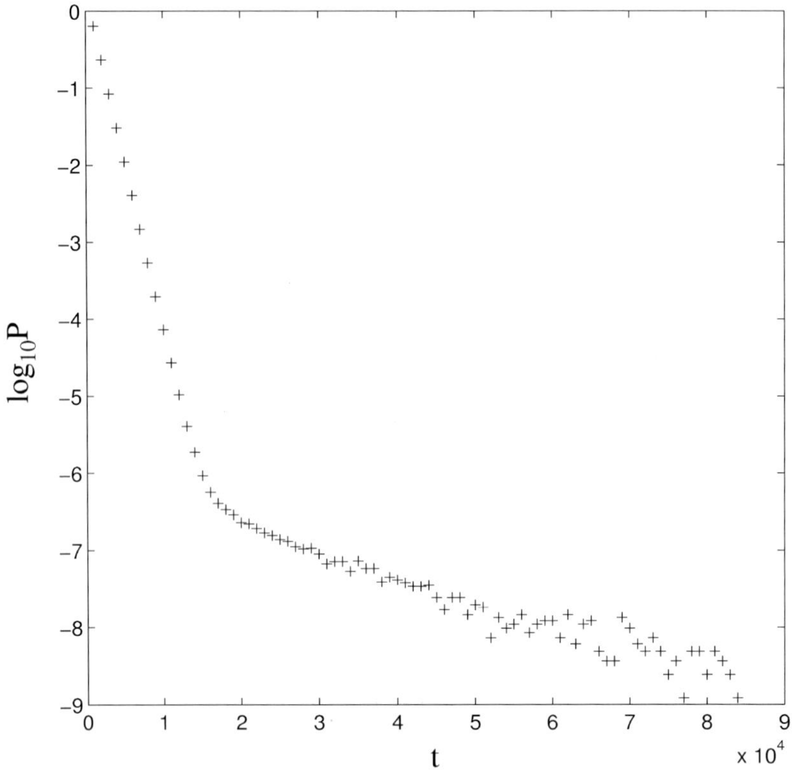

Figure 15.3. Distribution function of the Poincaré recurrences for the standard map in the situation of Figure 15.1.

Chapter 16

Chaos Synchronization

The studies of the phenomenon of chaos synchronization are usually based upon the analysis of the existence of transversely stable invariant manifold that contains an invariant set of trajectories corresponding to synchronous motions. In this chapter we present an approach that relays upon the notions of topological synchronization and the dimension for Poincaré recurrences. We show how the dimension of Poincaré recurrences may serve as an indicator for the onset of synchronized chaotic oscillations, capable of detecting the regimes of chaos synchronization characterized by the frequency ratio $p : q$.

16.1. Synchronization

It is well known that coupling between the dissipative dynamical systems with chaotic behavior can result in the onset of synchronized chaotic oscillations (see, for instance, [93] and references therein). In other words, a system

$$\begin{aligned} \dot{x} &= f(x) + cF(x, y, c), \\ \dot{y} &= g(y) + cG(x, y, c), \end{aligned} \tag{16.1}$$

(where $x \in \mathbb{R}^m$, $y \in \mathbb{R}^n$, and c is a coupling parameter) can behave in such a way that the x- and y-components of solution

$$\big(x(t, x_0, y_0),\ y(t, x_0, y_0)\big)$$

manifest some type of synchrony for $t \geqslant t_0 \gg 1$, independent of initial conditions (x_0, y_0) in a large region in $\mathbb{R}^{n+m}$.

The most simple type of synchronous chaotic behavior is the regime of identical synchronization. In this regime the solutions of the coupled subsystems (16.1) satisfy the following property

$$\lim_{t\to\infty} \big|x(t, x_0, y_0) - y(t, x_0, y_0)\big| = 0. \tag{16.2}$$

Of course, in order to achieve this type of behavior in the case $m = n$, the right-hand side of the system (16.1) should satisfy the identity

$$f(x) + cF(x, x, c) = g(x) + cG(x, x, c). \tag{16.3}$$

For example, it is so if $f(x) \equiv g(x)$ and $F(x, x, c) = G(x, x, c) \equiv 0$. It is easy to see that when the identity (16.3) holds the system (16.1) has the invariant manifold $x = y$, known as the synchronization manifold. When all invariant trajectories (associated with chaotic behavior) in this manifold are stable in the directions transversal to the manifold, the subsystems generate identical synchronous chaotic motions.

However, if coupled subsystems in (16.1) are nonidentical, then we cannot expect the validity of (16.2) and the notion of synchronization have to be treated differently. Different notions of chaos synchronization such as identical synchronization [56,94,122] stochastic synchronization [4], generalized synchronization [110, 74], asymptotic synchronization [62], phase synchronization [102], and others were introduced to point out significant features of the synchronization phenomenon. The present chapter is of the same spirit: we explore temporal characteristics of the synchronous chaotic trajectories and give a definition of synchronization based on the notion of Poincaré recurrences [13].

16.1.1. Periodic oscillations

Assume that the systems $\dot{x} = f(x)$ has a linearly stable limit cycle say L_1, with the period τ_1 and the system $\dot{y} = g(y)$ has linearly stable limit cycles L_2 with the period τ_2. Thus, the system (16.1), for $c = 0$, has the attracting torus $T_0 = L_1 \times L_2$. If the rotation number $\rho_0 = \tau_1/\tau_2$ is rational, then T_0 consists of periodic orbits. If ρ_0 is irrational then every orbit on T_0 is dense (on it). For $c \neq 0$ and small enough, there still exists an invariant attracting torus T_c in a neighborhood of T_0 [53]. Generally, for an open region in the parameter space, the system (16.1) has stable limit cycles. The synchronization regime corresponds to the existence of the stable limit cycle, say L_c, on the torus T_c. The Poincaré rotation number for these values of parameters is rational, say $m_0/n_0 \in \mathbb{Q}$ and it means that the closed curve L_c makes m_0 rotations along the generator L_1 of the torus T_0 and n_0 rotations along the other one. In terms of individual subsystems, we can describe the regime as follows. The orbit L_c corresponds to the solution $x = x_c(t)$, $y = y_c(t)$ of the system (16.1) where $(x_c(t), y_c(t))$ is a τ_c-periodic vector function. One can introduce polar coordinates (a_i, θ_i) in a neighborhood of L_i, $i = 1, 2$, such that θ_i is an angular coordinate along L_i and a_1 (a_2) is an *amplitude coordinate* transversal to $L_1 \in \mathbb{R}^m$ ($L_2 \in \mathbb{R}^n$). Then (for small values of c) the solution $(x_c(t), y_c(t))$ can be expressed in the new coordinates in the form

$$a_1 = a_1(t), \qquad \theta_1 = \omega_1 t + \alpha_1(t) \bmod \tau_c,$$

$$a_2 = a_2(t), \qquad \theta_2 = \omega_2 t + \alpha_2(t) \bmod \tau_c,$$

where $a_1, a_2, \alpha_1, \alpha_2$ are τ_c-periodic functions and $\omega_1/\omega_2 = n_0/m_0$. Stability of the limit cycle L_c ensures the regime of oscillations with the frequency relation

$\omega_1 m_0 = \omega_2 n_0$ for some domain in the parameters space. This domain is called synchronization zone.

For the sake of simplicity assume that $m_0 = 1$. If we may introduce (a, θ)-coordinates in such a way that a_1 and a_2 are constants, $\alpha_1 \equiv 0$, $\alpha_2 \equiv 0$, then at the instant of time $t = t_x = \tau_c/\omega_1$ we have $\theta_1(t_x) = \theta_1(0) \bmod \tau_c$ and $x_c(t_x) = x_c(0)$. However only at the moment $t = t_y = \tau_c/\omega_2 = n_0 t_x$, the second coordinate $y_c(t_y) = y_c(0)$. In other words, the "period" t_x of oscillations in the x-subspace can be different from the period of those in the y-subspace, and their ratio is given by

$$\frac{t_x}{t_y} = \frac{1}{n_0}.$$

The same as is true if $m_0 \neq 1$ and then

$$\frac{t_x}{t_y} = \frac{m_0}{n_0}. \tag{16.4}$$

Assume now that for some parameter values the system (16.1) has an attractor A_c containing infinitely many orbits, such that for $(x_0, y_0) \in A_c$ the projections $x(t, x_0, y_0)$ and $y(t, x_0, y_0)$ of the solution $(x, y)(t, x_0, y_0)$ onto the x-subspace and y-subspace behave similarly. In order to define rigorously this similarity, we have to be sure that (x_0, y_0) belongs to a periodic orbit. In this case something like the equality (16.4) holds and the number m_0/n_0 is independent on the choice of the periodic orbit in the attractor. However, if (x_0, y_0) belongs to an aperiodic orbits we should define some quantities which are similar to the periods of periodic orbits. Then, we again need to have something like the equality (16.4) for these quantities. We use Poincaré recurrences in the capacity of desired quantities, and follow the approach of previous chapters to compare the Poincaré recurrences for different subsystems.

16.2. Poincaré recurrences

Orbits in Hamiltonian systems and limiting orbits in dissipative systems possess the property of a repetition of their behavior in time. This repetition is expressed in terms of Poincaré recurrences.

Consider a dynamical system with continuous time $f^t : M \to M$, where $t \in \mathbb{R}_+$ and M is the phase space of the system which is assumed to be a complete metric space. Given an open set $U \subset M$ and a point $z \in U$, let the exit time $t_1(z, U)$ be defined as the following number: if $f^t z \in U$ for any $t \in \mathbb{R}_+$ then $t_1(z, U) = \infty$; if there is $t_0 \in \mathbb{R}_+$ such that $f^{t_0} z \notin U$, then

$$t_1(z, U) = \inf\{t_0 > 0:\ f^{t_0} z \notin U\}.$$

Figure 16.1. Bold intervals correspond to values of time for which a point belongs to U.

The set U is open. Therefore if there is $\bar{t} > t_1(z, U)$ such that $f^{\bar{t}}z \in U$, then there exists a maximal interval $(\alpha, \beta) \ni \bar{t}$ such that $f^t z \in U$ for any $t \in (\alpha, \beta)$ – see Figure 16.1. Let a "first" return time be defined by

$$t(z, U) = \begin{cases} 0, & \text{if } t_1(z, U) = \infty; \\ \inf \frac{\alpha+\beta}{2}, & \text{if } t_1(z, U) < \infty, \end{cases} \tag{16.5}$$

where the infimum is taken over all maximal intervals (α, β) such that $\alpha \geqslant t_1(z, U)$ and $f^t z \in U$ if $t \in (\alpha, \beta)$. In particular, $f^{t(z,U)}z \in U$ (of course, $t(z, U)$ may be equal to ∞).

In other words, if the orbit going through a point z comes back to the set U, then the point on this orbit spends an interval of time (α, β) in U before leaving the set again. We might choose any value of time inside (α, β) in the capacity of the Poincaré recurrence for the point z. It seems natural to specify the mean value $(\alpha + \beta)/2$ for that. Furthermore, the number of intervals with different α, β, corresponding to return times can be very large (surely infinite). We choose the interval corresponding to the first return times.

DEFINITION 16.1. The number $t(z, U)$ is said to be the Poincaré recurrence for the set U specified by the point z. The number

$$\tau(U) = \inf_{z \in U} t(z, U)$$

is called the Poincaré recurrence for the set U.

This definition is related to the repetition of the motion along orbits of dynamical systems. However, we are going to deal with the properties of the repetition along x (or the y)-components of the solution of a system of (16.1) type. We have to extend the definition of Poincaré recurrences to the case of coupled subsystems.

16.2.1. Poincaré recurrences for subsystems

To gain some insight let us first consider the following example of a periodically perturbed oscillator:

$$\ddot{z} + k\dot{z} + \phi(z) = a \sin\theta, \quad \dot{\theta} = 1, \tag{16.6}$$

where the nonlinearity $\phi(z)$ is of the Duffing-type. It is well known (see, for instance, [60]) that for same values of parameters the system (16.6) undergoes the period-doubling bifurcation, and has a stable 4π-periodic limit cycle, say L. For the system (16.6), the phase space is the direct product $X \times Y$, where

$$X = \{(z, \dot{z})\} \subset \mathbb{R}^2, \qquad Y = \{\theta, \mod 2\pi\} = S^1.$$

Let

$$\{z = z_0(t),\ \dot{z} = \dot{z}_0(t)\} \subset X, \qquad \{\theta = t \bmod 2\pi\} \subset S^1$$

be a solution corresponding to L. It is simple to understand that the curve $z = z_0(t)$, $\dot{z} = \dot{z_0}(t)$, $t \in [0, 4\pi]$, which is the projection of L onto X, might possess points of the self-intersection. At each of these points, say $(z_*, \dot{z_*})$ we have $z_* = z_0(t_1) = z_0(t_2)$, $\dot{z}_* = \dot{z}_0(t_1) = \dot{z}_0(t_2)$, $t_1 \neq t_2$, $t_{1,2} \in (0, 4\pi)$. If L is close to the limit cycle at the bifurcation moment, then such points have to exist by simple geometrical reasons. Evidently, if U_1 is a small neighborhood of $(z_*, \dot{z}_*)$, then the first return time to U_1, along the curve $z = z(t)$, $\dot{z} = \dot{z}(t)$ is much smaller than the first return time along this curve to a small neighborhood U_2 of a point p, $U_2 \not\ni (z_*, \dot{z_*})$.

The example shows that not all points on the projections of the attractor onto individual subspaces are responsible for the "right" Poincaré recurrences – bad points could exist, and we should take them into account, in order to give a good definition of Poincaré recurrences of individual subsystems.

Let X, Y be complete metric spaces (see, for example, system (16.1) where $X = \mathbb{R}^m$, $Y = \mathbb{R}^n$) and $f^t : X \times Y \to X \times Y$ be a dynamical system with $t \in \mathbb{R}_+$. (For the system (16.1) the evolution operator f^t is determined by solutions $(x, y)(t, x_0, y_0)$ going through initial points (x_0, y_0).) Let A be a subset of the phase space $X \times Y$ (say an attractor) and $A_1 = \pi_1 A \subset X$, $A_2 = \pi_2 A \subset Y$ be its images under natural projections to X and Y correspondingly (i.e., $\pi_1(x, y) = x$, $\pi_2(x, y) = y$ for any point $(x, y) \in X \times Y$).

We denote by

$$\begin{aligned} x(t, x_0, y_0) &= \pi_1 f^t(x_0, y_0), \\ y(t, x_0, y_0) &= \pi_2 f^t(x_0, y_0), \end{aligned} \tag{16.7}$$

the x- and y-coordinates of the orbit going through the initial point (x_0, y_0). Let U_i be an open set in A_i, $i = 1, 2$, and $x_0 \in A_1$, $y_0 \in A_2$. Denote by Y_{x_0} the set of all values of $y \in Y$ such that $(x_0, y) \in A$, i.e. $Y_{x_0} = \pi_2(\pi_1^{-1}(x_0) \cap A)$, the set of the y-coordinates of all π_1-preimages of the point x_0 belonging to A_1. Similarly, let $X_{y_0} = \pi_1(\pi_2^{-1}(y_0) \cap A)$, the set of all values of $x \in X$ such that $(x, y_0) \in A$. Assume that $x_0 \in U_1$ ($y_0 \in U_2$). Introduce a number $t_1(x_0, U_1)$ ($t_2(y_0, U_2)$) as follows:

(i) If $x(t, x_0, y_0) \in U_1$ for any $y \in Y_{x_0}$ and any value of $t \geqslant 0$ then $t_1(x_0, U_1) := \infty$. (Similarly, if $y(t, x, y_0) \in U_2$ for any $x \in X_{y_0}$ and any value of $t \geqslant 0$ then $t_2(y_0, U_2) := \infty$.)

(ii) If there exist $y \in Y_{x_0}$ and $t_0 = t_0(y)$ such that $x(t_0, x_0, y) \notin U_1$, then

$$t_1(x_0, U_1) := \inf_{y \in Y_{x_0}} \inf\{t_0 \mid x(t_0, x_0, y) \notin U_1\}.$$

(Similarly, if there exist $x \in X_{y_0}$ and $t_0 = t_0(x)$ such that $y(t_0, x, y_0) \notin U_2$, then

$$t_2(y_0, U_2) := \inf_{x \in X_{y_0}} \inf\{t_0 \mid y(t_0, x, y_0) \notin U_2\}.)$$

Roughly speaking, $t_{1,2}$ are exit times from $U_{1,2}$. Since the set U_1 (U_2) is open, if there exists $\bar{t} > t_1(x_0, U_1)$ $(\bar{t} > t_2(y_0, U_2))$ such that $x(\bar{t}, x_0, y) \in U_1$ for some $y \in Y_{x_0}$ $(y(\bar{t}, x, y_0) \in U_2$ for some $x \in X_{y_0})$, then there is a maximal interval $(\alpha, \beta) \ni \bar{t}$ such that $x(t, x_0, y) \in U_1$ for any $t \in (\alpha, \beta)$ $(y(t, x, y_0) \in U_2$ for any $t \in (\alpha, \beta))$. Set

$$t(x_0, U_1) := 0, \quad \text{if } t_1(x_0, U_1) = \infty,$$
$$t(x_0, U_1) := \inf_{y \in Y_{x_0}} \inf \frac{\alpha + \beta}{2}, \quad \text{if } t_1(x_0, U_1) < \infty,$$

where the second infimum is taken over all maximal interval (α, β) such that $\alpha \geqslant t_1(x_0, U_1)$ and $x(t, x_0, y) \in U_1$, if $t \in (\alpha, \beta)$, $y \in Y_{x_0}$. In particular, $x(t(x_0, U_1), y) \in U_1$ for some $y \in Y_{x_0}$, if $t(x_0, U_1) \neq \infty$. Similarly, introduce

$$t(y_0, U_2) := 0, \quad \text{if } t_2(y_0, U_2) = \infty,$$
$$t(y_0, U_2) := \inf_{x \in X_{y_0}} \inf \frac{\alpha + \beta}{2}, \quad \text{if } t_2(y_0, U_2) < \infty,$$

where the second infimum is taken over all maximal interval (α, β) such that $\alpha \geqslant t_2(y_0, U_2)$ and $y(t, x, y_0) \in U_2$ if $t \in (\alpha, \beta)$ for some $x \in X_{y_0}$.

DEFINITION 16.2.

(i) The number $t(x_0, U_1)$ is said to be the x-Poincaré recurrence for the set U_1 specified by the point x_0. The number

$$\tau_x(U_1) := \inf_{x_0 \in U_1} t(x_0, U_1) \tag{16.8}$$

is said to be the x-Poincaré recurrence for the set U_1.

(ii) The number $t(y_0, U_2)$ is said to be the y-Poincaré recurrence for the set U_2 specified by the point $y_0 \in U_2$. The number

$$\tau_y(U_2) := \inf_{y_0 \in U_2} t(y_0, U_2) \tag{16.9}$$

is said to be the y-Poincaré recurrence for the set U_2.

In Definitions 16.1 and 16.2, we take the infimum not only over all points in the open set but also over all possible "branches" going through the point in it. Roughly speaking, if $x(T_1, x_0, y_1) \in U_1$ and $x(T_2, x_0, y_2) \in U_1$ and $T_1 < T_2$, then we prefer T_1 as the Poincaré recurrence.

Thus, we defined quantities, which are similar, in some sense, to periods of periodic orbits and may now try to define what does mean "synchronization" in this framework.

Definitions 16.1 and 16.2 look too cumbersome, however, they are constructive; as we see below they allow us to calculate Poincaré recurrence in specific situations.

16.3. Topological synchronization

Now we are ready to give a definition of a kind of synchronization which is (more or less) natural to call "topological synchronization". This synchronization has to occur for a large set of initial conditions, belonging to the basin of an attractor. We may use any notion of the attractor we wish. Just for the sake of definiteness choose the following one. A compact set A in the phase space M of a dynamical system $f^t : M \to M$ is called an attractor if there exists an open set U such that $f^t\overline{U} \subset U$, $t > 0$, i.e., U is an absorbing region, and $A = \bigcap_{t>0} f^t U$ (see, for instance, [73,60,14]). Let us also note that a point $z \in M$ is non-wandering if $\tau(U) < \infty$ for any open set $U \ni z$ [73].

We give the definition first and below discuss each point of it.

DEFINITION 16.3. A dynamical system $f^t : X \times Y \to X \times Y$ is said to be (m_0/n_0)-topologically synchronized if:

(i) It has an attractor A such that nonwandering orbits are dense in A.
(ii) There is a number $N \in Z_+$ such that for any point $x_0 \in \pi_1(A)$, the set Y_{x_0} contains at most N points, and for any point $y_0 \in \pi_2(A)$ the set X_{y_0} contains at most N points.
(iii) There is a subset of "bad points" $B \subset A$ (B might be empty) such that if $A_1 = \pi_1(A)$, $A_2 = \pi_2(A)$, $B_1 = \pi_1(B)$, $B_2 = \pi_2(B)$, then

$$\overline{\dim}_B(B_i) < \dim_H(A_i), \quad i = 1, 2, \tag{16.10}$$

where $\overline{\dim}_B(\dim_H)$ is the upper box (Hausdorff) dimension.
(iv) For any point $(x_0, y_0) \in A \setminus B$, there are numbers ε_0, $a_1 \geqslant a_2 \geqslant 1$, such that: for any open set $U_1 \subset X$, $U_1 \ni x_0$, $\operatorname{diam} U_1 \leqslant \varepsilon \leqslant \varepsilon_0$, there is an open set $U_2 \subset Y$, $\operatorname{diam} U_2 \leqslant a_1(\operatorname{diam} U_1)$, $U_2 \ni y_0$, and for any open set $\widetilde{U}_2 \subset Y$, $\widetilde{U}_2 \ni y_0$, $\operatorname{diam} \widetilde{U}_2 \leqslant \varepsilon \leqslant \varepsilon_0$, there is an open set $\widetilde{U}_1$, $\operatorname{diam} \widetilde{U}_1 \leqslant a_2(\operatorname{diam} \widetilde{U}_2)$, $\widetilde{U}_1 \ni x_0$, such that

$$\tau_y(U_2) = \frac{m_0}{n_0}\tau_x(U_1) + \beta_2, \tag{16.11}$$

$$\tau_x(\widetilde{U}_1) = \frac{n_0}{m_0}\tau_y(\widetilde{U}_2) + \beta_1, \tag{16.12}$$

where $m_0, n_0 \in Z_+$, and $\beta_1 = \beta_1(\widetilde{U}_1, \widetilde{U}_2)$, $\beta_2 = \beta_2(U_1, U_2)$ are bounded as $\varepsilon \to 0$.

(v) If $\delta(B)$ is an open δ-neighborhood of the set B, where δ is small enough, then the constants ε_0, a_1, a_2 can be chosen to be the same for any point $(x_0, y_0) \in A \setminus \delta(B)$. They depend only on δ. Furthermore, the functions $\beta_{1,2}$ can be estimated from above by a constant $\bar{\beta} > 0$ depending only on δ and ε: $|\beta_{1,2}| \leqslant \bar{\beta}$.

(vi) If $A_{1\delta} = \pi_1(A \setminus \delta(B))$, $A_{2\delta} = \pi_2(A \setminus \delta(B))$ are subsets in A_1 and A_2 correspondingly which do not contain the set B_1 and B_2 of bad points together with some neighborhoods, then for any cover $G_1 = \{U_{1i}\}$ of the set $A_{1\delta}$ by open sets U_{1i} with $\operatorname{diam} U_{1i} \leqslant \varepsilon$, where $i \in I$, the finite or countable set of indices, there exists a cover $G_2 = \{U_{2j}\}$ of the set $A_{2\delta}$ by open sets of $\operatorname{diam} U_{2j} \leqslant a_2\varepsilon$, where $j \in J$, the finite or countable collection of indices, such that there is a map $\xi : J \to I, i = \xi(j)$, this map is onto, and

$$\operatorname{diam} U_{2j} \leqslant a_2 \operatorname{diam} U_{1\xi(j)} \tag{16.13}$$

for any $j \in J$. On Figure 16.2 the conditions (16.13) will not be satisfied in a neighborhood of bad points.

Moreover the number of preimages $\xi^{-1}(i)$ is finite and bounded from above by a constant S which is independent of the particular choice of the cover G_1 and of ε and depends only on δ: $S = S(\delta)$. Furthermore, for any fixed $i \in I$ and any $j \in J$ such that $\xi(j) = i$, the condition (16.12) is satisfied, i.e.

$$\tau_y(U_{2j}) = \frac{m_0}{n_0}\tau_x(U_{1i}) + \beta_2, \tag{16.14}$$

where $|\beta_2| \leqslant \bar{\beta} < \infty$, and the constant $\bar{\beta}$ depends only on δ.

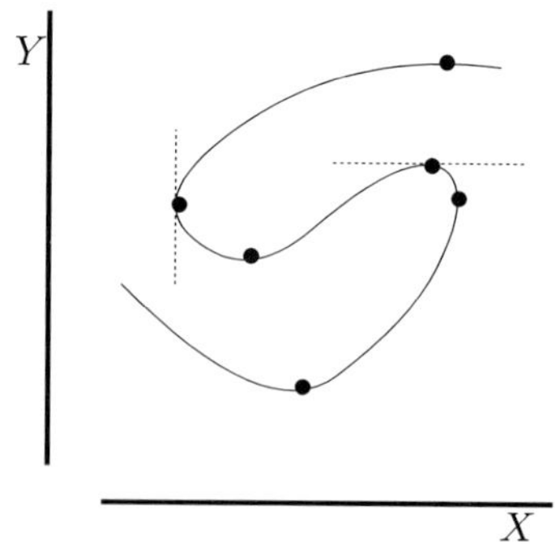

Figure 16.2. The curve is "an attractor" in $X \times Y$ and bold points are bad.

Similarly, for any cover $G_2 = \{\widetilde{U}_{2i}\}$ of the set $A_{2\delta}$ by open set $\widetilde{U}_{2i}$ with $\operatorname{diam} \widetilde{U}_{2i} \leqslant \varepsilon$, where $\{i\} = I$ is a finite or countable collection of indices, there exists a cover $\widetilde{U}_{1j}$ with $\operatorname{diam} \widetilde{U}_{1j} \leqslant a_1\varepsilon$, where $\{j\} = J$ is a finite or countable collection of indices, such that there is a map $\eta\colon J \to I,\ i = \eta(j)$, which is onto, and

$$\operatorname{diam} U_{1j} \leqslant a_1 \operatorname{diam} U_{2\eta(j)} \tag{16.15}$$

for any $j \in J$. The number of preimages $\eta^{-1}(i)$ is bounded from above by a constant S depending only on δ, and for any fixed $i \in I$ and $j \in J$, such that $\eta(j) = i$, we have

$$\tau_x\big(\widetilde{U}_{1j}\big) = \frac{n_0}{m_0}\tau_y\big(\widetilde{U}_{2\eta(j)}\big) + \beta_1, \tag{16.16}$$

where $|\beta_1| \leqslant \bar{\beta}$, and $\bar{\beta}$ is a constant depending only on δ.

REMARK 16.1.

(a) The condition (i) shows that Poincaré recurrences are finite for any open set, and, moreover, one should observe synchronization for $t \gg 1$ for open set of initial conditions.

(b) The condition (ii) claims that the projections π_1 and π_2 are finite-to-one maps. It is a natural assumption which is known to be satisfied, for example, if $f^t|_A$ is a minimal flow and coupling is unidirectional [33].

(c) The inequalities (16.10) mean that the "bad points" occupy a small part of the attractor.

(d) The condition (ii) implies also that if A_1 and A_2 contain infinitely many points then uncoupled subsystems can not be treated as synchronized ones. Indeed, for any point $x_0 \in A_1$ there are infinitely many (not at most N) points $\{y_0\} = A_2$ such that a solution through any of points (x_0, y_0) belong to an attractor.

(e) The assumption (v) implies that if the set $A \setminus \delta(B)$ contains periodic orbits of arbitrary large periods, then for infinitely many of them a relationship, similar to (16.4) holds. Indeed, let $L \subset A \setminus \delta(B)$ be a T-periodic orbit: $x = x(t, x_0, y_0),\ y = y(t, x_0, y_0)$, where x is a T_1-periodic function, y is a T_2-periodic function and at least one of the numbers $T_1,\ T_2$ equals T. Assume, for the sake of definiteness that $T_1 = T$. Then for an open ε-neighborhood U_1 of the point $x_0 \in X$ we have

$$\tau_x(U_1) = T_1 + \alpha_x(\varepsilon), \tag{16.17}$$

where $\alpha_x(\varepsilon) \to 0$ as $\varepsilon \to 0$. For the corresponding neighborhood U_2 of the point y_0 in Y with $\operatorname{diam} U_2 \leqslant a_1\varepsilon$ we have

$$\tau_y(U_2) = T_2 + \alpha_y(\varepsilon), \tag{16.18}$$

where $\alpha_y(\varepsilon) \to 0$ as $\varepsilon \to 0$. The relationship (16.12) can be rewritten as

$$T_2 + \alpha_y(\varepsilon) = \frac{m_0}{n_0} T + \frac{m_0}{n_0} \alpha_x(\varepsilon) + \beta_2,$$

i.e.,

$$\frac{T_2}{T} = \frac{m_0}{n_0} + \beta(\varepsilon) + \frac{\beta_2}{T},$$

where $\beta(\varepsilon) \to 0$ as $\varepsilon \to 0$ and $|\beta_2| \leqslant \bar{\beta}$. Thus,

$$\left| \frac{T_2}{T} - \frac{m_0}{n_0} \right| \leqslant \frac{\bar{\beta}}{T}.$$

Hence, if $T \gg 1$, then the ratio of periods in the x and y-subspaces is arbitrary close to $\frac{m_0}{n_0}$. We believe that generally it is $\frac{m_0}{n_0}$ exactly, but this conjecture has to be verified, of course.

(f) It is not a point to find a cover of the set $A_{2\delta}$, satisfying (16.13) if we know the cover of the set $A_{1\delta}$. For example, if points on A satisfy the equality $F(x, y) = 0$, F is a smooth vector function, and if the rank of the matrix $\frac{\partial F}{\partial y}$ is maximal and the rank of the matrix $\frac{\partial F}{\partial x}$ is maximal at points belonging to $A \setminus \delta(B)$, where B is the set of "critical points", then we have local diffeomorphism from a neighborhood of any point $x_0 \in A_{1\delta}$ to a neighborhood of a point $y_0 \in A_{2\delta}$ if $(x_0, y_0) \in A \setminus \delta(B)$. Therefore, the assumptions (16.13), (16.15) are automatically satisfied. However, the assumptions (16.14), (16.16) are satisfied only in the synchronized regimes. In order to recognize such regimes we need to use some characteristics of fractal dimension type.

16.4. Dimensions for Poincaré recurrences as indicators of synchronization

Consider the partition functions

$$M_x(\alpha, q, \varepsilon) = \inf_{G_1} \sum_i \exp\bigl(-q\tau_x(U_{1i})\bigr)(\operatorname{diam} U_{1i})^{\alpha}, \tag{16.19}$$

$$M_y(\alpha, q, \varepsilon) = \inf_{G_2} \sum_i \exp\bigl(-q\tau_y(U_{2i})\bigr)(\operatorname{diam} U_{2i})^{\alpha}, \tag{16.20}$$

where in each sum the infimum is taken over all covers G_1 (correspondingly G_2) of the set A_1 (correspondingly A_2) by open sets with diameters $\leqslant \varepsilon$.

DEFINITION 16.4.

(i) The critical values $\alpha_c^{(x)}(q)$ in (16.19) and $\alpha_c^{(y)}(q)$ in (16.20) are said to be spectra of dimensions for the x-Poincaré recurrences and, correspondingly, for the y-Poincaré recurrences.

(ii) If $\alpha_c^{(x)}(q_0^{(x)}) = 0$ (correspondingly $\alpha_c^{(y)}(q_0^{(y)}) = 0$), then the number $q_0^{(x)}$ (correspondingly $q_0^{(y)}$) is said to be the dimension for the x-Poincaré recurrences (for the y-Poincaré recurrences).

We notice first that under assumptions of Definition 16.3 the "individual attractors" A_1 and A_2 have the same Hausdorff dimensions.

THEOREM 16.1. (See [10].) *Assume that a dynamical system $f^t : X \times Y \to X \times Y$ is topologically synchronized (with respect to an attractor A). Then*

$$\dim_H(A_1) = \dim_H(A_2).$$

Remind that $\dim_H$ means the Hausdorff dimension.

Let us explain the main step of the proof. Let $\alpha_{1\delta} = \dim_H A_{1\delta}$, $\alpha_{2\delta} = \dim_H A_{2\delta}$, where $A_{i\delta} = \pi_i(A \setminus \delta(B))$, $i = 1, 2$.

Given $\alpha > \alpha_{2\delta}$, $K > 0$, consider a finite cover $\{\widetilde{U}_{2i}\}$, the set $A_{2\delta}$ by open sets with $\operatorname{diam} \widetilde{U}_{2i} \leqslant \varepsilon \leqslant \varepsilon_0$, such that

$$\sum_i (\operatorname{diam} \widetilde{U}_{2i})^\alpha \leqslant K, \tag{16.21}$$

such a cover exists, by the definition of the Hausdorff dimension. Consider the corresponding cover $G_1 = \{\widetilde{U}_{1j}\}$ of the set $A_{1\delta}$, $\operatorname{diam} \widetilde{U}_{1j} \leqslant a_1\varepsilon$, which exists thanks to Assumption (vi) in Definition 16.3. Considering (16.15), we have

$$\begin{aligned}\sum_j (\operatorname{diam} \widetilde{U}_{1j})^\alpha &= \sum_i \sum_{j,\eta(j)=i} (\operatorname{diam} \widetilde{U}_{1j})^\alpha \\ &\leqslant \sum_i \sum_{j,\eta(j)=i} a_1^\alpha (\operatorname{diam} U_{2\eta(j)})^\alpha \leqslant \sum_i a_1^\alpha S(\delta)(\operatorname{diam} U_{2i})^\alpha.\end{aligned}$$

The last inequality follows from Assumption (vi) in Definition 16.3. Hence due to (16.21),

$$\sum_j (\operatorname{diam} \widetilde{U}_{1j})^\alpha \leqslant a_1^\alpha S(\delta) K.$$

Since K is an arbitrary number, the relation above means that $\alpha > \dim_H(A_{1\delta})$ too, and therefore $\alpha_{2\delta} \geqslant \alpha_{1\delta}$. Similarly, we may start with a cover $\{U_{1j}\}$ of the set $A_{1\delta}$ and obtain the opposite inequality $\alpha_{1\delta} \geqslant \alpha_{2\delta}$. Thus, $\alpha_{1\delta} = \alpha_{2\delta}$. The remainder of the proof can be found in [10].

Thus, the theorem tells us that projections of the attractor on the individual subspaces at least have the same Hausdorff dimensions. Now we consider dimensions for the x- and y-Poincaré recurrences.

THEOREM 16.2. (See [10].) *If a dynamical system $f^t : X \times Y \to X \times Y$ is $\frac{m_0}{n_0}$-topologically synchronized, then*

$$q_0^{(y)}(A_2 \setminus B_2) = \frac{m_0}{n_0} q_0^{(x)}(A_1 \setminus B_1), \tag{16.22}$$

where q_0 is the dimension for Poincaré recurrences.

Let us show the main step of the proof. Consider again the sets $A_{1\delta}$ and $A_{2\delta}$ of "good points" in A_1 and A_2 correspondingly. Let $\alpha_c^{(x)}(q, A_{1\delta})$ ($\alpha_c^{(y)}(q, A_{2\delta})$) be the spectrum of dimensions for the x-Poincaré recurrences (for the y-Poincaré recurrences). Assume that $\alpha > \alpha_c^{(y)}(q, A_{2\delta})$. Then by definition of $\alpha_c^{(y)}(q, A_{2\delta})$, there is a cover $G_2 = \{\widetilde{U}_{2i}\}$, $\operatorname{diam} \widetilde{U}_{2i} \leqslant \varepsilon$, such that

$$\sum_i \exp(-q\tau_y(\widetilde{U}_{2i}))(\operatorname{diam} \widetilde{U}_{2i})^\alpha \leqslant K, \tag{16.23}$$

where K is a small number. Consider the corresponding cover $\{\widetilde{U}_{1j}\}$, $\operatorname{diam} \widetilde{U}_{1j} \leqslant a_1\varepsilon$, of the set $A_{1\delta}$, satisfying (16.16). Then

$$\begin{aligned}
&\sum_j \exp\left(-q\frac{m_0}{n_0}\tau_x(\widetilde{U}_{1j})\right)(\operatorname{diam} \widetilde{U}_{1j})^\alpha \\
&\quad = \sum_i \sum_{j,\eta(j)=i} \exp\left(-q\frac{m_0}{n_0}\tau_x(\widetilde{U}_{1j})\right)(\operatorname{diam} \widetilde{U}_{1j})^\alpha \\
&\quad \leqslant \sum_i \sum_{j,\eta(j)=i} \exp\left(-q\frac{m_0}{n_0}\tau_x(\widetilde{U}_{1j})\right)a_1^\alpha(\operatorname{diam} \widetilde{U}_{2j})^\alpha =: Q.
\end{aligned}$$

Thanks to (16.16), the last expression Q can be estimated as follows

$$\begin{aligned}
Q &\leqslant S(\delta)a_1^\alpha e^{q\frac{m_0}{n_0}\bar{\beta}} \sum_i \exp(-q\tau_y(\widetilde{U}_{2i}))(\operatorname{diam} \widetilde{U}_{2i})^\alpha \\
&\leqslant KS(\delta)a_1^\alpha \exp\left(q\frac{m_0}{n_0}\bar{\beta}\right).
\end{aligned}$$

The last inequality follows from (16.23). Hence, this α, which is greater than $\alpha_c^{(y)}(q, A_{2\delta})$, satisfies the inequality

$$\alpha > \alpha_c^{(x)}\left(\frac{m_0}{n_0}q, A_{1\delta}\right).$$

Therefore,

$$\alpha_c^{(x)}\left(\frac{m_0}{n_0}q, A_{1\delta}\right) \leqslant \alpha_c^{(y)}(q, A_{2\delta}).$$

Starting with a cover $\{U_{1i}\}$ of the set $A_{1\delta}$ and repeating the proof above, we obtain that

$$\alpha_c^{(y)}\left(\frac{n_0}{m_0}q, A_{2\delta}\right) \leqslant \alpha_c^{(x)}(q, A_{1\delta}). \tag{16.24}$$

Now assume that $q_{0i} := \dim_P(A_{i\delta})$, i.e., $\alpha_c^{(x)}(q_{01}) = 0$, $\alpha_c^{(y)}(q_{02}) = 0$. Since $\alpha_c^{(x)}$ and $\alpha_c^{(y)}$ are monotone, then (16.24) implies that $\alpha_c^{(y)}(\frac{n_0}{m_0}q, A_{2\delta}) \leqslant 0$ and therefore $q_{02} \leqslant \frac{m_0}{n_0}q_{01}$. Similarly, $q_{01} \leqslant \frac{n_0}{m_0}q_{02}$. It follows that $q_{02} = \frac{m_0}{n_0}q_{01}$.

REMARK 16.2. We believe that (under some general conditions),

$$q_0^{(y)}(A_2) = \frac{m_0}{n_0}q_0^{(x)}(A_1),$$

as well. Of course, the Poincaré recurrences on the "bad sets" B_1 and B_2 can be different from those on the $A_1 \setminus B_1$ and $A_2 \setminus B_2$. However, since

$$\overline{\dim}_B B_i < \dim_H(A_i \setminus B_i), \quad i = 1, 2,$$

by assumption, we believe that a "randomly chosen" point on A_i belongs to $A_i \setminus B_i$. In numerical simulations we may neglect "bad points", if they exist, and treat the equality (16.22) as an indicator of $\frac{m_0}{n_0}$-synchronization. In other words, if subsystems are $\frac{m_0}{n_0}$-synchronized, then

$$\langle\tau_x(U_{1i})\rangle \sim -\frac{b}{q_0^{(x)}(A_1)}\ln\varepsilon, \tag{16.25}$$

$$\langle\tau_y(U_{2i})\rangle \sim -\frac{b}{q_0^{(y)}(A_2)}\ln\varepsilon, \tag{16.26}$$

and

$$q_0^{(y)}(A_2) = \frac{m_0}{n_0}q_0^{(x)}(A_1).$$

16.5. Computation of Poincaré recurrences

In this section we consider an algorithm to compute the average of the x- and y-Poincaré recurrences $\langle\tau_x(U_{1i})\rangle$ and $\langle\tau_y(U_{2i})\rangle$ as in (16.8) (16.9), respectively, with various diameters ε for the open covers. We compute the average of $\tau_x(U_{1i})$ and $\tau_y(U_{2i})$ by

$$\langle\tau_x(U_{1i})\rangle = \frac{1}{N_x}\sum_i \tau_x(U_{1i}), \qquad \langle\tau_y(U_{2i})\rangle = \frac{1}{N_y}\sum_i \tau_y(U_{2i}).$$

Here N_x (N_y) is the number of elements U_{1i} (U_{2i}) with diam $U_{1i} \leqslant \varepsilon$ (diam $U_{2i} \leqslant \varepsilon$) in the cover of the set A_1 (A_2).

As it was mentioned in Remark 16.2, we expect that the averages $\tau_x(U_{1i})$ and $\tau_y(U_{2i})$ of topologically synchronized subsystems behave as

$$\langle \tau_x(U_{1i}) \rangle \sim -\frac{b}{q_0^{(x)}(A_1)} \ln \varepsilon, \qquad \langle \tau_y(U_{2i}) \rangle \sim -\frac{b}{q_0^{(y)}(A_2)} \ln \varepsilon. \tag{16.27}$$

To study the Poincaré recurrence of the continuous-time dynamical system $f^t : X \times Y \to X \times Y$, we first integrate the system of differential equations with arbitrary given initial value and determine an invariant set $A \subset X \times Y$. Next we consider open coverings for $A_1 = \pi_1(A) \subset X$ and $A_2 = \pi_2(A) \subset Y$. For the invariant sets A_1 and A_2 we define sets of open balls $\{B(x_i), \varepsilon\} \equiv \{U_{1i}^{\varepsilon}\}$ and $\{B(y_i), \varepsilon\} \equiv \{U_{2i}^{\varepsilon}\}$ of radius ε which uniformly cover A_1 and A_2 respectively. We use these sets of balls to compute the average of the Poincaré recurrence $\langle \tau_x(U_{1i}^{\varepsilon}) \rangle$. From (16.8) and (16.9) we know that $\tau_x(U_{1i}^{\varepsilon})$ is the infimum of the first return time $t(x_0, U_{1i}^{\varepsilon})$ over all x_0 in U_{1i}^{ε}. For each open ball we can compute the first return time and then we calculate the average of the Poincaré recurrence $\langle \tau_x(U_{1i}^{\varepsilon}) \rangle$. Then we change ε and proceed with the same analysis to compute another value of $\langle \tau_x(U_{1i}^{\varepsilon}) \rangle$. For the analysis of the dimension for Poincaré recurrences we plot $\langle \tau_x(U_{1i}^{\varepsilon}) \rangle$ against $-\ln \varepsilon$. Similarly, we also compute $\langle \tau_y(U_{2i}^{\varepsilon}) \rangle$ and plot the graph $\langle \tau_y(U_{2i}^{\varepsilon}) \rangle$ against $-\ln \varepsilon$.

In the remaining part of this section we present a few examples which show how the regimes of chaos synchronization of systems having different individual dynamics can be studied by means of the Poincaré recurrence.

EXAMPLE 16.1. *Mutually coupled nonlinear oscillators with parametric excitation.* The synchronization of chaotic oscillations in such oscillators was studied in [4]. The dynamics of the coupled parametric oscillators is given by the following differential equations

$$\begin{cases} \dfrac{dx_1}{dt} = x_2, \\ \dfrac{dx_2}{dt} = -k_1 x_2 - x_1(1 + q_1 \cos \Omega t + x_1^2) - c(x_2 - y_2), \end{cases}$$

$$\begin{cases} \dfrac{dy_1}{dt} = y_2, \\ \dfrac{dy_2}{dt} = -k_2 y_2 - y_1(1 + q_2 \cos \Omega t + y_1^2) + c(x_2 - y_2), \end{cases}$$

where parameters are $k_1 = 0.48,\ k_2 = 0.45,\ q_1 = q_2 = 50,\ \Omega = 2$.

Figure 16.3 presents attractors and plots of Poincaré recurrences calculated for the case of uncoupled oscillators. Since $k_1 \neq k_2$ the uncoupled oscillators ($c = 0$)

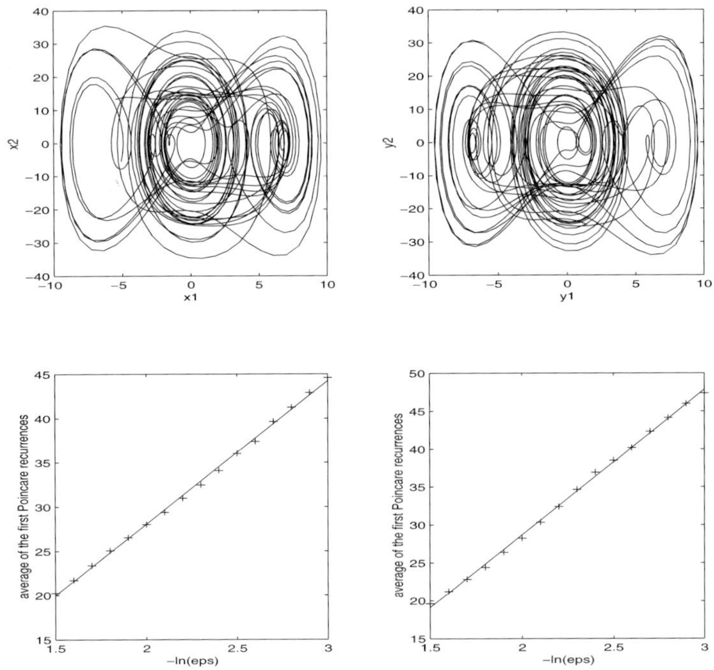

Figure 16.3. Uncoupled parametric oscillators. Phase portraits of the chaotic attractors (top) and plots of the corresponding Poincaré recurrences (bottom).

have different dynamics. As a result, the plots $\langle\tau(U_i^\varepsilon)\rangle$ versus $(-\ln\varepsilon)$, calculated for the attractors in the phase spaces x and y, have different slopes 12.9 and 19.2, respectively. The dispersion of the calculated values of the slopes is about 2%. Therefore difference of the slopes is quite significant, which indicates that attractors have different dimensions for Poincaré recurrences.

When the parametric oscillators are synchronized the plots of $\langle\tau(U_i^\varepsilon)\rangle$ versus $(-\ln\varepsilon)$, calculated for the 'x' and 'y' oscillators, have the slopes 19.05 and 19.07, respectively. These slopes are the same taking into account accuracy of our calculations.

Figure 16.4 presents such plots and attractors calculated for the systems with the coupling parameter $c = 80$.

EXAMPLE 16.2. *Mutually coupled Lorenz systems.* They are modeled by the system of differential equations

$$\begin{cases} \dot{x}_1 = \sigma_1(x_2 - x_1) + c_1(y_1 - x_1), \\ \dot{x}_2 = \rho_1 x_1 - x_2 - x_1 x_3 + c_2(y_2 - x_2), \\ \dot{x}_3 = -\beta_1 x_3 + x_1 x_2 + c_3(y_3 - x_3), \end{cases}$$

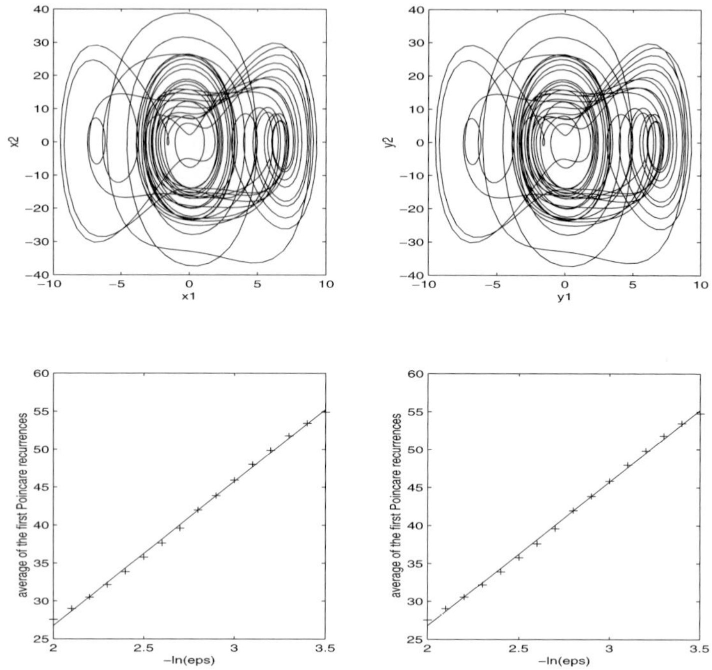

Figure 16.4. Synchronized parametric oscillators. Phase portraits of the chaotic attractors (top) and plots of the corresponding Poincaré recurrences (bottom).

$$\begin{cases} \dot{y}_1 = \sigma_2(y_2 - y_1) + c_1(x_1 - y_1), \\ \dot{y}_2 = \rho_2 y_1 - y_2 - y_1 y_3 + c_2(x_2 - y_2), \\ \dot{y}_3 = -\beta_2 y_3 + y_1 y_2 + c_3(x_3 - y_3). \end{cases}$$

Here we consider a case of slightly nonidentical systems. The parameters are chosen as follows: $\sigma_1 = 16.0$, $\sigma_2 = 16.02$, $\rho_1 = 45.92$, $\rho_2 = 45.92$, $\beta_1 = 4.0$, $\beta_2 = 4.01$.

The results of numerical simulations of Lorentz systems and calculations Poincaré recurrences for the case of uncoupled systems ($c_1 = c_2 = c_3 = 0$) and for the case of synchronized system (with $c_1 = 500$, $c_2 = c_3 = 400$) are presented in Figures 16.5 and 16.6, respectively.

16.6. Final remarks

In this chapter we have introduced a notion of topological synchronization of coupled chaotic subsystems. Roughly speaking, two subsystems are topologically

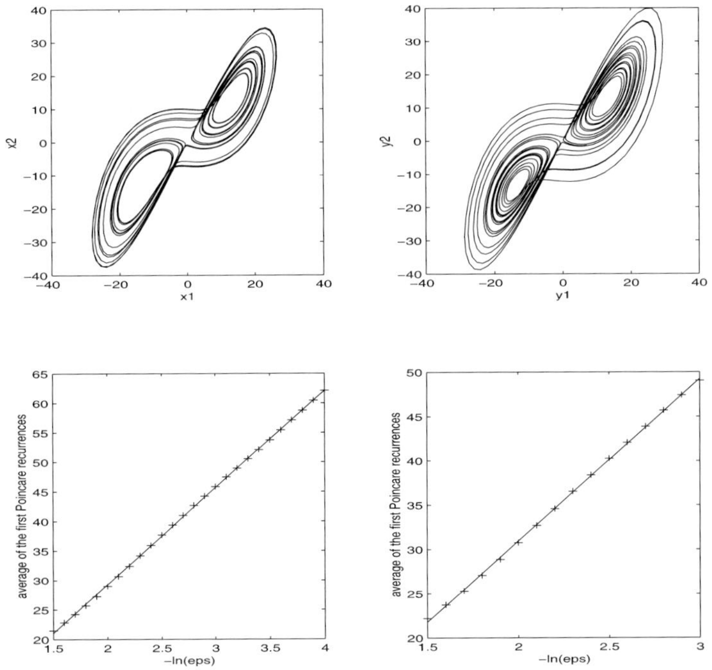

Figure 16.5. Uncoupled Lorenz systems. Phase portraits of the chaotic attractors (top) and plots of the corresponding Poincaré recurrences (bottom). The slopes calculated for the plots of the Poincaré recurrences of 'x' and 'y' systems are 16.54 and 18.43 respectively.

synchronized if their Poincaré recurrences behave similarly. It means that the first return time to a small spot of initial states in the first subsystem is approximately the same as the corresponding one for the second subsystem, provided that sizes of spots are about the same. In more general case, the ratio of first return times for both systems is approximately a rational number which does not dependent on the positions of spots in attractors. An indicator of similarity of behavior is based on the notion of the dimension for Poincaré recurrences. Coincidence (or, more generally, rational ratio) of dimensions for two subsystems is an only necessary condition for topological synchronization, since it only shows that a similarity occurs "on average": the mean of first return times to small spots for the first subsystem is approximately the same as the one for the second subsystem. In principal, there is a logical possibility that Poincaré recurrences behave differently in different parts of attractors, while mean values of them are approximately the same. Nevertheless, we believe that the criterion we suggested is useful and may serve well in specific situations with nonidentical subsystems. We would like to

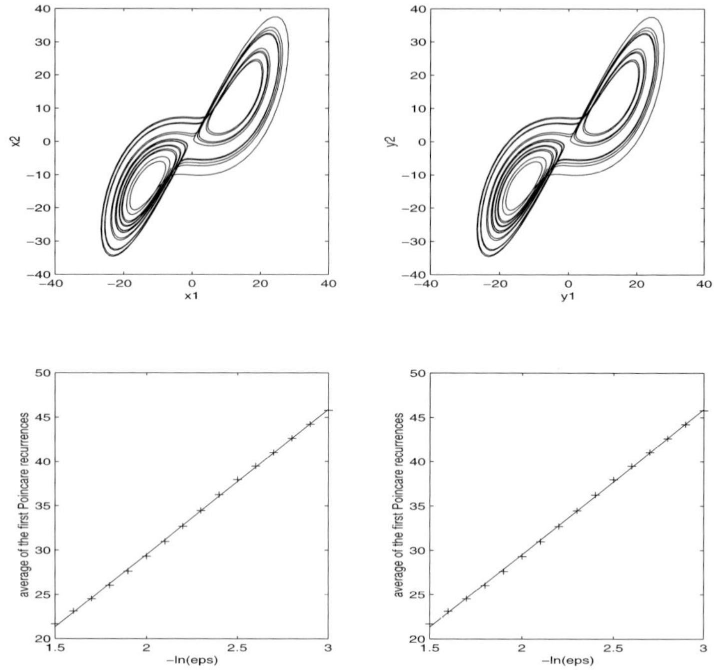

Figure 16.6. Synchronized Lorenz systems. Phase portraits of the chaotic attractors (top) and plots of the corresponding Poincaré recurrences (bottom). The slopes calculated for the plots of the Poincaré recurrences of 'x' and 'y' systems are the same and equal to 16.27.

emphasize that for the case of mutual coupling of nonidentical chaotic subsystems there are no sufficiently general criteria of synchronized regimes.

As it was mentioned in Introduction a very important feature of synchronization phenomenon is that a particular frequency relation, or, in our case, a relation between the Poincaré recurrences, does not change within the synchronization zone. The invariance of such relations is the essence of synchronization regime. Indeed, two uncoupled identical chaotic oscillators have the same characteristics of Poincaré recurrences. However, this is only a result of identity of these oscillators. There is no synchronization between them because they are uncoupled and produce oscillations that are not correlated to each other. In this case a small mismatch of parameters in these oscillators will change the relation between their Poincaré recurrences. Being properly coupled the oscillators synchronize and the established relation between their Poincaré recurrences does not change with the parameters mismatch while the parameters are in the synchronization zone.

Here, we also would like to discuss briefly the relation between the asymptotic equality for mean values of the exponents of Poincaré recurrences and mean value for Poincaré recurrences themselves. In Section 4.3 (see also Definition 16.4), the asymptotic equality

$$\left\langle e^{-q_0\tau(U_i)}\right\rangle \sim \varepsilon^b$$

shows that we may expect that

$$\left\langle \exp\left(-q_0^{(x)}\tau_x(U_{1i})\right)\right\rangle \sim \varepsilon^{b_1},$$
$$\left\langle \exp\left(-q_0^{(y)}\tau_y(U_{2i})\right)\right\rangle \sim \varepsilon^{b_2},$$

where $q_0^{(x)} = \dim_{\mathrm{p}}(A_1)$, $q_0^{(y)} = \dim_{\mathrm{p}}(A_2)$ and $b_i = \dim_{\mathrm{B}}(A_i)$, $i = 1, 2$. We may also expect that $\dim_B(A_i) = \dim_H(A_i)$, $i = 1, 2$. In this case, Theorem 16.1 implies the asymptotic equalities

$$\left\langle \tau_x(U_{1i})\right\rangle \sim -\frac{b}{q_0^{(x)}(A_1)}\ln\varepsilon \quad \text{and} \quad \left\langle \tau_y(U_{2i})\right\rangle \sim -\frac{b}{q_0^{(y)}(A_2)}\ln\varepsilon$$

as in (16.27), where $b = b_1 = b_2$ and $\{U_{1i}\}$, $\{U_{2i}\}$ are open coverings of A_1 and A_2 by open balls of diameter ε, designed as in Section 16.5. Thus, the approximation (16.27) serves indeed as a basic indicator of synchronized regimes.

The numerical results we presented here are related to the cases of (1:1)-synchronization only. However some preliminary considerations show that the criterion works for $(p{:}q)$-topological synchronization as well.

PART VI

APPENDICES

Chapter 17

Some Known Facts about Recurrences

Let (X, μ, f) be a dynamical system with X a compact set. First we shall see that for any subset $A \subset X$, $\mu(A) > 0$, almost every $x \in A$ returns infinitely many times to A. In such a situation, the natural statistical quantity of interest is the average first return time to A, $E(r_A)$. Then we will prove Kac's theorem that implies that, for ergodic μ, $E(r_A) = \mu(A)^{-1}$. A similar quantity is the average of the return times of the orbit of $x \in A$ to A, $t_A(x)$. In the case μ is an ergodic measure we will see that $t_A(x) = E(r_A)$ for almost every x in A.

17.1. Almost everyone comes back

In this section we prove Poincaré recurrence theorem.

Let $A \subset X$ have $\mu(A) > 0$. The set A is not necessarily invariant. The return time to A of point $x \in A$ is the number

$$r_A(x) := \min\{n > 0\colon\ f^n(x) \in A\}.$$

Let $A_n \subset A$ denote the set of all points in A that return to it for the first time after n time-steps,

$$\begin{aligned} A_n &:= \{x \in A\colon\ r_A(x) = n\} \\ &= \{x \in A\colon\ f^n(x) \in A \text{ and } f(x), f^2(x), \ldots, f^{n-1}(x) \notin A\}. \end{aligned}$$

We have that $A_n \subset A \cap f^{-n}(A)$ and the set $A_1 \subset A$ contains the points that eventually get off A. The set A_1 includes the non wandering subset $\bigcap_{n\geqslant 0} f^n(A_1)$ that is invariant and might be nonempty. If μ is ergodic, then $\mu(\bigcap_{n\geqslant 0} f^n(A_1)) = 0$ or 1 since $f^{-1}(\bigcap_{n\geqslant 0} f^n(A_1)) = \bigcap_{n\geqslant 0} f^n(A_1)$. We will encounter only situations where almost every point in A eventually leaves set A: what comes in will eventually go out.

The subset of points that leave set A to never come back to it is denoted by

$$A_\infty = A \setminus \bigcup_{n>0} \{x \in A\colon\ f^n(x) \in A\},$$

$$= A \setminus \bigcup_{n>0} A \cap f^{-n}(A),$$

$$= \bigcap_{n>0} A \setminus f^{-n}(A).$$

Thus, A_∞ is measurable. The collection $\{A_n\colon n > 0\}$ together with A_∞ constitute a measurable disjoint partition of the set A.

Let $x \in A_\infty \subset A$. Then, by definition, $f^k(x) \notin A_\infty$ for every $k > 0$, i.e., $f^{-k}(A_\infty) \cap A_\infty = \emptyset$ for $k > 0$. Thus,

$$f^{-m}\big(f^{-k}(A_\infty) \cap A_\infty\big) = f^{-k-m}(A_\infty) \cap f^{-m}(A_\infty) = \emptyset \quad \forall k, m > 0.$$

We formulate this fact in the form of the following.

LEMMA 17.1. *$f^{-k}(A_\infty) \cap f^{-\ell}(A_\infty) = \emptyset$ whenever $k \neq \ell$.*

PROPOSITION 17.1. *For any invariant probability measure μ, $\mu(A_\infty) = 0$.*

PROOF. By Lemma 17.1 we have that

$$\mu\Bigg(\bigcup_{k>0} f^{-k}(A_\infty)\Bigg) = \sum_{k>0} \mu\big(f^{-k}(A_\infty)\big).$$

By invariance of μ we have that $\mu(\bigcup_{k>0} f^{-k}(A_\infty)) = \sum_{k>0} \mu(A_\infty) < \infty$. Thus $\mu(A_\infty) = 0$. □

Hence, the collection $\{A_n\colon n > 0\}$ is a disjoint partition of almost all points in A, relative to any invariant measure μ.

THEOREM 17.1. (Poincaré, 1912.) *Let $\mu(A) > 0$. Then, for almost every $x \in A$ there exists $n = r_A(x) > 0$ such that $f^n(x) \in A$.*

PROOF. By Proposition 17.1 we have that $\mu(\bigcup_{n>0} A_n) = \mu(A)$. □

Of course, Theorem 17.1 implies that a.e. $x \in A$ returns infinitely many times to A. Indeed, the Poincaré's theorem applies to each f^k, $k > 0$, and for each k: $\mu(A_\infty^{(k)}) = 0$. Thus, $A \setminus \bigcup_{k>0} A_\infty^{(k)}$ has full measure and a.e. $x \in A \setminus \bigcup_{k>0} A_\infty^{(k)}$ returns infinitely many times to A.

A direct consequence of Poincaré's theorem is that for a.e. $x \in A$ the sequence $\sum_{k=0}^{n-1} \mathbb{1}_A(f^k(x))$ diverges as $n \to \infty$. More generally, for $h \in L_1(X, \mu)$ a non-negative function, $0 \neq h \geqslant 0$, the sequence $\sum_{k=0}^{n-1} h(f^k(x))$ diverges for a.e. $x \in \{y\colon h(y) > 0\}$. The proof is easy for simple functions, which constitute a set that is everywhere dense in $L_1(X, \mu)$.

17.2. Kac's theorem

Let $W_n = (X \setminus A) \cap f^{-n}(A)$ denote the set of points in $X \setminus A$ that hit set A after n steps for the first time,

$$W_n := \{x \notin A\colon\ f(x), \dots, f^{n-1}(x) \notin A \text{ and } f^n(x) \in A\}, \quad n > 0.$$

The set of points in $X \setminus A$ that never hit the set A is denoted by W_∞. When the invariant measure is not ergodic the set of points in $X \setminus A$ that never reach set A may have positive measure, i.e., $\mu(W_\infty) \neq 0$, generally. For convenience we let $W_0 := A$.

LEMMA 17.2. *The following relations hold.*

(1) *For* $n > 0,\ f(A_n) \subset W_{n-1}$.
(2) $f^{-1}(A) = A_1 \cup W_1$.
(3) $f^{-1}(W_n) = A_{n+1} \cup W_{n+1}$.

Transitions induced by f between the sets A_n and W_n specify a transition matrix that is depicted in Figure 17.1. Nodes of the graph denote sets A_k and W_k. The transition graph has an arrow from the node denoting set B to the node denoting set C whenever $f(B) \cap C \neq \emptyset$. The transition graph is usually represented as the tower in Figure 17.2, known as the Rokhlin tower of f induced in A. However, the tower representation might be misleading since the return to A, the transition from W_1 to A, is not necessarily to the subset A_n that was left originally by the point. The transition is from W_1 to any of the A_k. As drawn in the tower Figure 17.2 the dynamics may be confused with a collection of uncoupled cycles. And it is not so. We like graph (Figure 17.1) better than tower (Figure 17.2).

THEOREM 17.2. (Kac, 1947 [71], see also [70].) *For invariant (not necessarily ergodic)* μ*:*

$$\int_A r_A\, d\mu = \sum_{k>0} k\mu(A_k) = 1 - \mu(W_\infty).$$

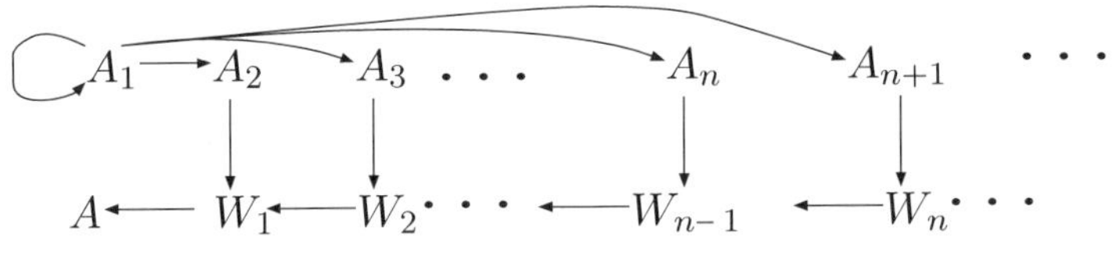

Figure 17.1. Transition graph of f induced in $A \subset X$.

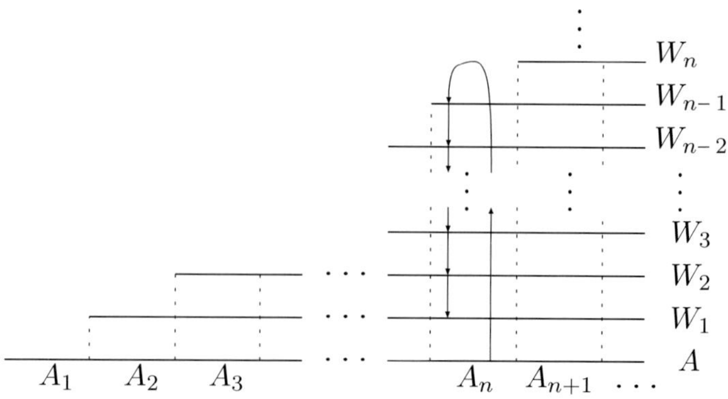

Figure 17.2. Rokhlin tower of f induced in $A \subset X$.

PROOF. The collection of sets $\{W_n\} \cup \{A_n\} \cup \{W_\infty,\ A_\infty\}$ constitutes a disjoint partition of X, so that

$$1 = \mu(X) = \sum_{n>0} \mu(A_n) + \sum_{n>0} \mu(W_n) + \mu(W_\infty), \tag{17.1}$$

since $\mu(A_\infty) = 0$. By property (3) in Lemma 17.2 we have that $\mu(W_n) = \mu(f^{-1}(W_n)) = \mu(A_{n+1}) + \mu(W_{n+1})$ and then (see the transition graph Figure 17.1)

$$\mu(W_n) = \sum_{k>0} \mu(A_{n+k}) \tag{17.2}$$

since $\mu(W_n) \to 0$ as $n \to \infty$ (this is so because $\sum_{n>0} \mu(W_n) < \infty$). Substituting result (17.2) in (17.1) the theorem follows. □

After Kac's theorem, we know that in a dynamical system (X, f, μ) (μ is invariant but not necessarily ergodic) the natural time scale associated to set $A \subset X$, $\mu(A) > 0$, is its average return time (or the Poincaré cycle of A),

$$\bar{\tau}(A) := \frac{1}{\mu(A)} \int_A r_A \, d\mu = \frac{\mu(X \setminus W_\infty)}{\mu(A)}.$$

The usual situation is that of an ergodic measure μ for which $\mu(X \setminus W_\infty) = \mu(X) = 1$. However, in the study of volume-preserving maps (e.g., Chapter 15) we deal with Lebesgue measure which is not ergodic. See [85] for a discussion of Kac's theorem for volume preserving maps.

Chapter 18

Birkhoff's Individual Theorem

Given a dynamical system (X, f, μ) and a function $h \in L_1(X, \mu)$, the aim in this Appendix is to prove the point, almost everywhere, convergence of the sequence

$$\frac{1}{n}\sum_{k=0}^{n-1} h\big(f^k(x)\big)$$

to a function $\bar{h}(x) \in L_1(X, \mu)$. The limit $\bar{h}(x)$ is the average of h along the forward orbit of the point x. To prove Birkhoff's theorem we need to prove the Hopf maximal ergodic theorem first, which looks as a generalization of the Chebyshev inequality.

18.1. Some general definitions

Life is easier if everything is expressed in terms of so called doubly stochastic operators. So, let us introduce some convenient definitions. Given function f, Koopman's operator F (an isometry, see property (2)) on $L_1(X, \mu)$ is defined by $Fh = h \circ f$, for every $h \in L_1(X, \mu)$. F is a doubly stochastic operator since it satisfies the following properties,

(1) $h \geqslant 0$ (a.e.) $\Longrightarrow Fh \geqslant 0$ (a.e.),
(2) $\int_X Fh\, d\mu = \int_X h\, d\mu$,
(3) $F1 = 1$ (a.e.),

where 1 denotes the function whose constant value is 1 (a.e.). When (X, μ) is an interval with μ the Lebesgue measure, the doubly stochastic operators on $L_1(X, \mu)$ (i.e., those verifying conditions (1)–(3) above) are just exactly those operators that are defined by a map $f : X \to X$ (measurable and preserving μ) [34].

The positive, h^+, and negative, h^-, parts of function h are defined to be

$$h^+(x) = \begin{cases} 0, & h(x) \leqslant 0, \\ h(x), & \text{otherwise} \end{cases}$$

and $h^- = (-h)^+$. Thus, $h = h^+ - h^-$, $|h| = h^+ + h^-$ and $h^+, h^- \geqslant 0$.

18.2. Proof of the Birkhoff's theorem

Let each $F_n h : X \to \mathbb{R}$, $n > 0$, denote the following partial sums by

$$F_n h(x) := \sum_{j=0}^{n-1} F^j h(x) = \sum_{j=0}^{n-1} h\big(f^j(x)\big)$$

(be warned that $F_1 h = h \neq Fh$). For the set of partial sums $\{F_k h\colon k = 1, \dots, n\}$ let us define its "roof" at a point x to be the number

$$h_n(x) := \max_{1 \leqslant k \leqslant n} F_k h(x).$$

It should be evident that $h_n \leqslant h_{n+1}$. The following lemma puts an upper bound to the rate of growing of h_n.

LEMMA 18.1. $h_{n+1} \leqslant h + Fh_n^+$.

PROOF. For given n we are going to prove that

$$F_k h \leqslant h + Fh_n^+ \quad \text{for } k = 1, \dots, n+1.$$

Since F is non-negative, for $k = 1$, it follows, by the definition of $F_n h$, that

$$F_1 h = h \leqslant h + Fh_n^+.$$

For $k + 1$,

$$\begin{aligned} F_{k+1} h &= \sum_{j=0}^{k} F^j h = h + \sum_{j=1}^{k} F^j h = h + F \sum_{j=0}^{k-1} F^j h \\ &= h + F(F_k h). \end{aligned}$$

Since F is non-negative and $h_k \leqslant h_n$, $k \leqslant n$, we have that

$$F(F_k h) \leqslant Fh_n \leqslant Fh_n^+, \qquad \text{for } k = 1, \dots, n.$$

Thus, $F_{k+1} h \leqslant h + Fh_n^+$ for $k = 1, \dots, n$. □

Consider next the set of points where the first n partial sums have a positive roof,

$$\{h_n > 0\} := \big\{x \in X\colon h_n(x) > 0\big\}, \qquad n > 0.$$

Remark that $\{h_n > 0\}$ is the set of points x such that there exists a $k \in \{1, \dots, n\}$ for which $F_k(x) > 0$. The sequence of sets is nondecreasing: $\{h_n > 0\} \subset \{h_{n+1} > 0\}$.

Let us further define

$$B_*(h) := \bigcup_{n>0} \{h_n > 0\},$$

the set of points $x \in X$ where, for some n, the partial sum $F_n h(x)$ is positive.

The following maximal ergodic theorem is due to Hopf. We don't have any reference to the original source. In [63], Halmos gives a reference to Hopf which is dated 1937. Thus, the following theorem seems to be younger than Birkhoff's.

THEOREM 18.1. (Hopf, $\leqslant$ 1937.) *For each $h \in L_1(X, \mu)$, $\int_{B_*(h)} h\, d\mu \geqslant 0$.*

PROOF. By Lemma 18.1 we have that $h \geqslant h_n - Fh_n^+$ (remember that $h_n \leqslant h_{n+1}$). Thus

$$\int_{\{h_n>0\}} h\, d\mu \geqslant \int_{\{h_n>0\}} h_n\, d\mu - \int_{\{h_n>0\}} Fh_n^+\, d\mu.$$

Since $h_n|\{h_n > 0\} = h_n^+|\{h_n > 0\}$ and $h_n^+|\{h_n > 0\}^c = 0$. Then,

$$\int_{\{h_n>0\}} h\, d\mu \geqslant \int_X h_n^+\, d\mu - \int_X Fh_n^+\, d\mu = 0.$$

The equality holds because F is doubly stochastic and satisfies property (2). Letting $n \to \infty$ the theorem follows. □

Regarding the sequence of partial sums $F_n h(x)$ let us consider the following quantities,

$$h^*(x) := \sup_n \frac{1}{n} F_n h(x), \qquad h_*(x) := \inf_n \frac{1}{n} F_n h(x),$$

$$\bar{h}(x) := \limsup_{n\to\infty} \frac{1}{n} F_n h(x) \quad \text{and} \quad \underline{h}(x) := \liminf_{n\to\infty} \frac{1}{n} F_n h(x).$$

It should be evident that $h_* \leqslant \underline{h} \leqslant \bar{h} \leqslant h^*$. Moreover, $\{\bar{h} > 0\} \subset \{h^* > 0\}$ and $\{\bar{h} > 0\} = B_*(h)$.

For a given $\alpha \in \mathbb{R}$ consider the following subsets of X:

$$\{h^* > \alpha\} := \{x\colon h^*(x) > \alpha\} \quad \text{and} \quad \{h_* < \alpha\} := \{x\colon h_*(x) < \alpha\}.$$

COROLLARY 18.1. *For each $h \in L_1(X, \mu)$ and each $\alpha \in \mathbb{R}$ we have*

$$\alpha\mu\{h^* > \alpha\} \leqslant \int_{\{h^*>\alpha\}} h\, d\mu \quad \textit{and} \quad \alpha\mu\{h_* < \alpha\} \geqslant \int_{\{h_*<\alpha)\}} h\, d\mu.$$

PROOF. Observe that $F_n(h(x) - \alpha) > 0$ iff $F_n h(x)/n > \alpha$. Then, the first inequality is obtained by applying the Hopf Theorem 18.1 to the function $h - \alpha \in L_1(X, \mu)$. The second inequality is not independent of the first one. It is obtained from the first inequality by applying it to $-h$ and then replacing $\alpha \mapsto -\alpha$. □

There are similarities of above corollary with Chebyshev inequality. For instance, if we take F to be the identity, $Fh = h$, then $h^* = h_* = h$ and the first inequality is Chebyshev's in [75], p. 341. Similar situation if F is not the identity but $h \in L_1(X, \mu)$ is F-invariant: $Fh = h$ (a.e.) (remember that F is an isometry). For simple random variables see [26], p. 276.

THEOREM 18.2. (Birkhoff individual ergodic theorem, 1931.) *Let (X, f, μ) be a dynamical system and let $h \in L_1(X, \mu)$. Then*

$$\lim_{n\to\infty} \frac{1}{n} \sum_{k=0}^{n-1} h\big(f^k(x)\big) = \bar{h}(x) = \underline{h}(x) \quad (a.e.).$$

PROOF. We are going to prove that the set of points

$$\{\underline{h} < \bar{h}\} := \big\{x \in X: \underline{h}(x) < \bar{h}(x)\big\},$$

where the functions $\underline{h}$ and $\bar{h}$ take different values, has zero measure: $\mu\{\underline{h} < \bar{h}\} = 0$.

For each pair of rational numbers $\beta < \alpha \in \mathbb{Q}$, consider the set

$$A(\alpha, \beta) := \big\{x \in X: \underline{h}(x) < \beta < \alpha < \bar{h}(x)\big\} = \{\underline{h} < \beta\} \cap \{\bar{h} > \alpha\}.$$

We are going to prove that $\mu(A(\alpha, \beta)) = 0$. Assume the contrary: there exist numbers $\beta < \alpha \in \mathbb{Q}$ such that $\gamma := \mu(A(\alpha, \beta)) > 0$. Then, since $\bar{h}(f(x)) = \bar{h}(x)$ and $\underline{h}(f(x)) = \underline{h}(x)$, it is clear that the two-level sets are invariant: $f(A(\alpha, \beta)) \subset A(\alpha, \beta)$. Thus, we may consider the dynamical system $(A(\alpha, \beta), f|A(\alpha, \beta), \mu_{A(\alpha,\beta)})$ and apply Corollary 18.1 to it. Let us then remark that in this case $X = A(\alpha, \beta)$ and

$$\{h^* > \alpha\} = \big\{x \in A(\alpha, \beta): h^*(x) > \alpha\big\} \subset A(\alpha, \beta).$$

Since $\bar{h} \leqslant h^*$, then $\bar{h}(x) > \alpha \Longrightarrow h^*(x) > \alpha$ for every $x \in A(\alpha, \beta)$. Thus $\{h^* > \alpha\} = A(\alpha, \beta)$. Quite similarly

$$\{h_* < \beta\} = \big\{x \in A(\alpha, \beta): h_*(x) < \beta\big\} = A(\alpha, \beta),$$

since $h_* \leqslant \underline{h}$ and then $\underline{h}(x) < \beta \Longrightarrow h_*(x) < \beta$ for every $x \in A(\alpha, \beta)$. Inequalities in Corollary 18.1 are then

$$\alpha \leqslant \frac{1}{\gamma} \int_{A(\alpha,\beta)} h \, d\mu \leqslant \beta$$

which contradict the fact that $\beta < \alpha$.

Moreover, $\mu\{\underline{h} < \bar{h}\} = 0$. It follows from the fact that

$$\{\underline{h} < \bar{h}\} = \bigcup_{\substack{\beta<\alpha \\ \alpha,\beta\in\mathbb{Q}}} \{\underline{h} < \beta\} \cap \{\bar{h} > \alpha\},$$

since then $\mu\{\underline{h} < \bar{h}\} \leqslant \sum_{\beta<\alpha\in\mathbb{Q}} \mu(A(\alpha, \beta)) = 0$.

To prove that $\bar{h}$ is integrable we just verify conditions in the Fatou's lemma (see [75], pp. 346–347). We already have almost everywhere convergence of $F_n h/n$. Then, we are left to prove that there exists K such that $\int_X F_n h/n \, d\mu \leqslant K$ for every n. That it is so is shown in the following expressions,

$$\int_X \left| \frac{1}{n} F_n h \right| d\mu \leqslant \frac{1}{n} \int_X \sum_{k=0}^{n-1} \left| h\big(f^k(x)\big) \right| d\mu(x) = \int_X |h| \, d\mu.$$

Moreover (by the Fatou's lemma too)

$$\int_X |\bar{h}| \, d\mu \leqslant \int_X |h| \, d\mu. \qquad \square$$

Chapter 19

The Shannon–McMillan–Breiman Theorem

19.1. Introduction

In its more general version, the Shannon–McMillan–Breiman theorem establishes the μ-a.e. existence the limit

$$h := \lim_{n\to\infty} \frac{-\log \mu(\zeta^n(x))}{\#A_n},$$

where

(a) $X \ni x$ is a measurable space;
(b) $\mathcal{G} \supset A_n$ is a semigroup of measurable transformation in Ω;
(c) μ is a $\mathcal{G}$-ergodic measure;
(d) ζ is a finite measurable partition of X, and for each $n \in \mathbb{N}$, $\zeta^n(x)$ denotes the atom in

$$\zeta^n := \bigvee_{g\in A_n} g^{-1}\zeta$$

containing x;
(e) the sequence A_n satisfies
 (i) $\lim_{n\to\infty}(gA_n \Delta A_n)/\#A_n = 0$, for all $g \in \mathcal{G}$,
 (ii) $A_n \subset A_{n+1}$ for all $n \in \mathbb{N}$,
 (iii) there exists $M > 0$ such that $\#(A_n^{-1} \cap A_n) \leqslant M\#A_n$ for all $n \in \mathbb{N}$.

This general version can be found in [91].

The version of the theorem we use in the book refers to the case where the semigroup of transformations is one-dimensional, generated by the iterations of a transformation $T : \Omega \to \Omega$. The traditional proof of this theorem, due to L. Breiman (see [32]), which is based on the Martingale Convergence Theorem. Here we presented a more elementary proof relaying only on the Birkhoff's ergodic theorem, which is an adaptation to the one dimensional case of the Ornstein and Weiss' proof of the general version mentioned above.

19.2. The theorem

The framework will be the same as in Section 3.1, i.e., (X, T) is a subshift of Ω_p, we consider in X a Borel probability measure μ, which we suppose ergodic with respect to the shift transformation. Let ζ be the partition of X by 1-cylinders, and for each $n \in \mathbb{N}$

$$\zeta^n := \bigvee_{k=0}^{n-1} T^{-k}\zeta,$$

and for each $x \in X$, $\zeta^n(x)$ denotes the cylinder of length n containing x. The entropy of μ with respect to the shift is the limit

$$h(\mu) := -\lim_{n\to\infty} \frac{1}{n} \sum_{\zeta^n(x)\in\zeta^n} \mu\bigl(\zeta^n(x)\bigr) \log \mu\bigl(\zeta^n(x)\bigr),$$

whose existence is ensure by subadditivity, and it is always finite for a subshift $X \subset \Omega_p$.

THEOREM 19.1. (Shannon–McMillan–Breiman.)

$$\mu\left\{x \in X\colon \lim_{n\to\infty} \frac{-\log \mu(\zeta^n(x))}{n} = h(\mu)\right\} = 1.$$

19.3. Proof of the theorem

The proof of this theorem may be divided into two parts. Let us first present the simplest one.

Lower bound. For each $x \in X$ and $n \in \mathbb{N}$ let $h_n := -\log \mu(\zeta^n(x))/n$, then define the function $h\colon X \to [0, \infty]$, with $h(x) = \liminf_{n\to\infty} h_n(x)$. We first prove that $h(Tx) \leqslant h(x)$ in a set of full measure μ. For this, fix $\varepsilon > 0$ and for each $n \in \mathbb{N}$ let

$$E_n^\varepsilon := \bigl\{x \in X\colon h_n(Tx) \geqslant h_n(x) + \varepsilon\bigr\}.$$

Since $\zeta^n(Tx) \supset \zeta^{n+1}(x)$, then

$$\frac{\mu(\zeta^{n+1}(x))}{\mu(\zeta^n(x))} \leqslant \frac{\mu(\zeta^n(Tx))}{\mu(\zeta^n(x))} \leqslant \frac{e^{-nh_n(Tx)}}{e^{-nh_n(x)}} \leqslant e^{-\varepsilon n},$$

for each $x \in E_n^\varepsilon$. This implies that for $c \in \zeta^n$ and $\tilde{c} \in \zeta^{n+1}$ with $\tilde{c} \subset c$ and $\tilde{c} \cap E_n^\varepsilon \neq \emptyset$, we have $\mu(\tilde{c}) \leqslant e^{-\varepsilon n}\mu(c)$. Now, since each cylinder of length n

contains at most p cylinders of length $n+1$, then

$$\mu\big(c \cap E_n^\varepsilon\big) \leqslant pe^{-\varepsilon n}\mu(c)$$

for each $c \in \zeta^n$, therefore

$$\mu\big(E_n^\varepsilon\big) = \sum_{c\in\zeta^n} \mu\big(c \cap E_n^\varepsilon\big) \leqslant pe^{-\varepsilon n} \sum_{c\in\zeta^n} \mu(c) = pe^{-\varepsilon n}.$$

With this we have proved that for each $\varepsilon > 0$, $\sum_{n=1}^\infty \mu(E_n^\varepsilon) < \infty$, hence, by the Borel–Cantelli Lemma we have

$$\mu\big\{x \in X\colon \exists n_0 \text{ such that } x \notin E_n^\varepsilon\ \forall n \geqslant n_0\big\} = 1.$$

Therefore for each $\varepsilon > 0$ there exists a set $X_\varepsilon \subset X$ of full μ measure such that $h(Tx) < h(x) + \varepsilon$ for each $x \in X_\varepsilon$. Then, the set $\widetilde{X} = \bigcap_{k=1}^\infty X_{1/k}$ is such that $\mu(\widetilde{X}) = 1$ and $h(Tx) \leqslant h(x)$ for each $x \in \widetilde{X}$.

Now, since h is sub-invariant almost everywhere, then it has to be invariant almost everywhere. Indeed, let $\widetilde{Y} := \bigcap_{n=0}^\infty T^n\widetilde{X}$, and for each $\varepsilon > 0$ consider the measurable set $\widetilde{Y}_\varepsilon := \{x \in \widetilde{X}\colon h(Tx) \leqslant h(x) - \varepsilon\}$. If for each $x \in \widetilde{Y}_\varepsilon$ there exists $n = n(x)$ such that $T^k(x) \neq \widetilde{Y}_\varepsilon$ for each $k \geqslant n$, it means that $\widetilde{Y}_\varepsilon$ is not recurrent, and the Poincaré recurrence theorem implies that $\mu(\widetilde{Y}_\varepsilon) = 0$. This being true for arbitrary ε implies that $\mu\{x \in \widetilde{Y}\colon h(Tx) = h(x)\} = \mu(\widetilde{Y}) = 1$. Finally, for μ ergodic, $h(x)$ is constant almost everywhere.

Summarizing, there exists a constant $h \geqslant 0$ such that

$$h(x) := \liminf_{n\to\infty} \frac{-\log \mu(\zeta^n(x))}{n} = h,$$

in a set of full measure μ.

Upper bound. Fix $\varepsilon > 0$ and for each $n \in \mathbb{N}$ define

$$\mathcal{C}_n^\varepsilon := \left\{c \in \zeta^n\colon \frac{-\log \mu(c)}{n} \leqslant h + \varepsilon\right\}.$$

The lower bound we just proved implies that for all x in a set $X_{\mathcal{C}}$ of full measure, $\zeta^n(x) \in \mathcal{C}_n^\varepsilon$ for infinitely many n's.

Fix $n_0 \in \mathbb{N}$, and for $n \geqslant n_0$ define

$$T_{n_0}^n := \left\{\begin{array}{l} [x_0^{n-1}] \in \zeta^n\colon \exists\, a_1 < b_1 < \cdots < a_\ell < b_\ell \text{ satisfying} \\ \qquad\qquad \text{(a) } b_i - a_i + 1 := n_i \geqslant n_0, \\ \qquad\qquad \text{(b) } [x_{a_i}^{b_i}] \in \mathcal{C}_{n_i}, \\ \qquad\qquad \text{(c) } \sum_{i=1}^\ell n_i \geqslant (1 - 2/n_0)n. \end{array}\right\}.$$

Step 1. We first prove that for arbitrary n_0, and for all x in a set $Y_{\mathcal{C}}$ of full measure, there exists $m_0 > n_0$ such that $\zeta^n(x) \in T_{n_0}^n$ for all $n \geqslant m_0$.

For all $x \in X_{\mathcal{C}}$ we have $\zeta^n(x) \in \mathcal{C}_n^{\varepsilon}$ for infinitely many n's, therefore the measurable function

$$x \mapsto t(x) := \min\{k \geqslant n_0 \colon \zeta^k(x) \in \mathcal{C}_k\}$$

is finite in $X_{\mathcal{C}}$. From this it follows that

$$\mu\{x \in X_{\mathcal{C}} \colon t(x) > M\} < \frac{1}{4n_0}$$

for all positive integer M sufficiently large.

Fix such M, and let $B := \{x \in X_{\mathcal{C}} \colon t(x) > M\}$. Then, by Birkhoff's theorem

$$\mu\left\{x \in X_{\mathcal{C}} \colon \lim_{m\to\infty} \frac{1}{m} \sum_{j=0}^{m-1} \chi_B(T^j x) < \frac{1}{3n_0}\right\} = \mu(X_{\mathcal{C}}) = 1.$$

Thus, there exists a set $Y_{\mathcal{C}} \subset X_{\mathcal{C}}$ with $\mu(Y_{\mathcal{C}}) = \mu(X_{\mathcal{C}}) = 1$, and for each $x \in Y_{\mathcal{C}}$ there exists $m = m(x)$ such that for each $n \geqslant m$

$$\#\{0 \leqslant j < n - M \colon t(T^j x) \leqslant M\} \geqslant n\left(1 - \frac{1}{3n_0}\right) - M.$$

To the each index $k \in \{0 \leqslant j < n \colon t(T^j x) \leqslant M\}$ we associate the interval $[k, \ell(k)] \subset [0, n-1]$ with $\ell(k) = k + t(T^k x)$. Hence, we have a collection $S_n(x)$ of intervals of length in $[n_0, M]$, whose left ends cover a set of at least $n(1 - 1/(3n_0)) - M$ points in $[0, n-1]$. Hence, there exists a subcollection $\widetilde{S}_n(x) \subset S_n(x)$, composed by disjoint intervals, which covers a at least $n(1 - 1/(3n_0)) - M - n/(3n_0)$ points in $[0, n-1]$. Now, by taking $m_0 \geqslant m$ so large that $M \leqslant n/(3n_0)$ for all $n \geqslant m_0$, and with then defining a_i, b_i such that $\widetilde{S}_n(x) := \{[a_i, b_i] \colon i = 1, \ldots, \ell\}$, we obtain that for all $x \in Y_{\mathcal{C}}$ there exists $m_0 > n_0$ such that $\zeta^n(x) \in T_{n_0}^n$ for all $n \geqslant m_0$.

Step 2. Now, given n and n_0 as above, the number of collections

$$\widetilde{S}_n := \{[a_i, b_i] \colon i = 1, \ldots, \ell\}$$

such that $b_i - a_i + 1 \geqslant n_0$, and $\sum_{i=1}^{\ell} n_i \geqslant (1 - 2/n_0)n$, is bounded by the number of ways of choosing the left ends a_i of the intervals in $\widetilde{S}_n$. Since the total number of intervals in a collection $\widetilde{S}_n$ cannot exceed n/n_0, then the total number of collections is bounded by

$$\sum_{k=0}^{n/n_0} \binom{n}{k} \leqslant n\sqrt{n} \exp(ns(1/n_0)) =: N_1,$$

with $s(1/n_0) = -1/n_0 \log(1/n_0) - (1 - 1/n_0)\log(1 - 1/n_0)$. The last estimate follows in the standard way from Stirling's approximation.

Given a collection $\widetilde{S}_n$ such that $b_i - a_i + 1 \geqslant n_0$, and $\sum_{i=1}^{\ell} n_i \geqslant (1 - 2/n_0)n$, a cylinders $[x_0^{n-1}] \in T_{n_0}^n$ is said to by compatible with $\widetilde{S}_n$ if $\widetilde{S}_n(x) \equiv \widetilde{S}_n$. Thus, if $[x_0^{n-1}]$ is compatible with $\widetilde{S}_n := \{[a_i, b_i]\colon i = 1, \ldots, \ell\}$, then $x_{a_i}^{b_i}$ is one amongst at most $\exp(n_i(h+\varepsilon))$ possibilities. This is because by definition $[x_{a_i}^{b_i}] \in \mathcal{C}_{n_i}$, i.e., $[x_{a_i}^{b_i}]$ belongs to a set of cylinders of length n_i, each of them with measure bounded below by $\exp(-n_i(h + \varepsilon))$. In this way, the number of cylinders $[x_0^{n-1}] \in T_{n_0}^n$ compatible with a given collection $\widetilde{S}_n := \{[a_i, b_i]\colon i = 1, \ldots, \ell\}$ is bounded by

$$\left(\prod_{i=1}^{\ell} e^{n_i(h+\varepsilon)}\right) p^{n/n_0} \leqslant \exp\left(n(h+\varepsilon) + \frac{n}{n_0}\big(\log(p) - h - \varepsilon\big)\right) =: N_2,$$

which does not depend on $\widetilde{S}_n$.

From bounds N_1 and N_2 we obtain

$$\#T_{n_0}^n \leqslant N_1 \times N_2 \leqslant e^{n(h+\varepsilon - s(1/n_0) + \log(p)/n_0 + 3/2\log(n)/n)}.$$

Then, for n_0 sufficiently large we have $\#T_{n_0}^n \leqslant e^{n(h+2\varepsilon)}$.

Last Step. For n_0, n and ε as before, define

$$U_n := \left\{c \in T_{n_0}^n\colon \mu(c) \leqslant e^{-n(h+3\varepsilon)}\right\},$$

Then, $\mu(\bigcup_{c\in U_n} c) \leqslant \#T_{n_0}^n e^{-n(h+3\varepsilon)} \leqslant e^{-n\varepsilon}$. Here, applying again Borel–Cantelli we derive,

$$\mu\left\{x \in \widetilde{Y}\colon \zeta^n(x) \in U_n \text{ for infinitely many } n\text{'s}\right\} = 0.$$

Summarizing, given $\varepsilon > 0$ and $n_0 \in \mathbb{N}$, for all $x \in \widetilde{Y}$ there exists $m_0 = m_0(x)$ such that

(i) $\zeta^n(x) \in T_{n_0}^n$ for all $n \geqslant m_0(x)$,
(ii) $\zeta^n(x) \in T_{n_0}^n$ and $\mu(\zeta^n(x)) \leqslant e^{-n(h+3\varepsilon)}$ holds only for finitely many n's.

From this it follows that $\limsup_{n\to\infty} -\log\mu(\zeta^n(x))/n < h + 3\varepsilon$, for all $x \in \widetilde{Y}$, and since $\varepsilon > 0$ was taken arbitrary, then the upper bound follows.

Chapter 20

Amalgamation and Fragmentation

We construct homeomorphisms between minimal multipermutative systems satisfying conditions of Theorem 8.2. By using Theorem 7.1, we can assume that they are adding machines. We apply a technique of amalgamation and fragmentation of symbols (used, for instance, in [65] in the general context of minimal Cantor systems). The technique become easier in our context. Let us recall some definitions.

Let $q_0, q_1, \dots, q_n$ be positive integers and for $i = 0, \dots, n$ consider the alphabets $A_{q_i} := \{0, \dots, q_i - 1\}$ and $A_{q_0 \cdot q_1 \cdot \dots \cdot q_n} := \{0, \dots, q_0 \cdot q_1 \cdot \dots \cdot q_n - 1\}$, with the usual order. There is a natural lexicographical order induced on $A_{q_0} \times \dots \times A_{q_n}$. The map $\varphi_{n+1} : A_{q_0} \times \dots \times A_{q_n} \to A_{q_0 \cdot q_1 \cdot \dots \cdot q_n}$ that preserves the order is said to be an amalgamation of symbols. The inverse map is called a fragmentation of symbols. Both maps can be extended to infinite sequences of positive integers as follows.

For $q_* = (q_0, q_1, \dots)$ let $\Omega_{q_*} = \bigotimes_{j=0}^{\infty} A_{q_j}$. Let $m = (m_0 = 0, m_1, m_2, \dots)$ be an increasing sequence of integers. For every $i \geqslant 0$, define $q_i(m) = \prod_{j=m_i}^{m_{i+1}-1} q_j$ and $B_i(m) = \bigotimes_{j=m_i}^{m_{i+1}-1} A_{q_j}$. Finally, let $\Omega_{q_*}(m) = \bigotimes_{j=0}^{\infty} B_j(m)$ and $q_*(m) = \prod_{j=0}^{\infty} q_j(m)$. We say that $q_*(m)$ (correspondingly $\Omega_{q_*}(m)$) is produced from q_* (correspondingly Ω_{q_*}) by a m-amalgamation. Conversely, we say that q_* (correspondingly Ω_{q_*}) is produced from $q_*(m)$ (correspondingly $\Omega_{q_*}(m)$) by a m-fragmentation.

The map $\varphi(m) : \Omega_{q_*} \to \Omega_{q_*}(m)$ corresponding to a m-amalgamation is defined by the formula

$$\begin{aligned}
&\varphi(m)\,(\omega_0, \dots, \omega_{m_1-1}, \omega_{m_1}, \dots, \omega_{m_2-1}, \dots) \\
&\quad = \big(\varphi_{m_1}(\omega_0, \dots, \omega_{m_1-1}), \varphi_{m_2-m_1}(\omega_{m_1}, \dots, \omega_{m_2-1}), \dots\big),
\end{aligned}$$

for every $\omega \in \Omega_{q_*}$.

The definition implies that the map $\varphi(m)$ conserves the number $\#\,(p, q_*)$, therefore, thanks to Theorem 8.2, the adding machines (Ω_{q_*}, S) and $(\Omega_{q_*}(m), S(m))$ are topologically conjugate. In fact, $\varphi(m)$ is a conjugacy. Indeed, the map $\varphi(m)$ is one-to-one, and $S(m) \circ \varphi(m) = \varphi(m) \circ S$. Obviously, $\varphi(m)$ is continuous, hence, it is a homeomorphism.

Now we will describe a recursive procedure which allows one to pass from a polyadic adding machine (Ω_{q_*}, S) to a conjugate one $(\Omega_{q'_*}, S')$ by means of m-amalgamations and m-fragmentations.

Let $q_* = (q_0, q_1, \ldots)$ and $q'_* = (q'_0, q'_1, \ldots)$ be two equivalent sequences. We begin by defining recursively a third equivalent sequence $q''_* = (q''_0, q''_1, \ldots)$:

(1) Put $q''_0 = q_0$ and $m_0 = 1$.
(2) Let m_1 be the least integer such that q_0 divides $q'_0 \cdots q'_{m_1-1}$. It exists because sequences are equivalent. Define $q''_1 = q'_0 \cdots q'_{m_1-1}/q_0$.
(3) Now we define the recurrence. Assume we have already defined $q''_0, \ldots, q''_n$ and $m_0, \ldots, m_n$ for $n > 0$. If $n + 1$ is even, let m_{n+1} be the least integer greater than m_{n-1} such that q''_n divides $q_{m_{n-1}} \cdots q_{m_{n+1}-1}$ and define $q''_{n+1} = q_{m_{n-1}} \cdots q_{m_{n+1}}/q''_n$. If $n + 1$ is odd, let m_{n+1} be the least integer greater than m_{n-1} such that q''_n divides $q'_{m_{n-1}} \cdots q'_{m_{n+1}-1}$ and define $q''_{n+1} = q'_{m_{n-1}} \cdots q'_{m_{n+1}}/q''_n$.

The sequence q''_* satisfies the following properties (they follow directly from the definition):

(1) The following equalities hold

$$q''_{2n-1}q''_{2n} = q_{m_{2n-2}}q_{m_{2n-2}+1} \cdots q_{m_{2n}-1}, \qquad n = 1, 2, \ldots$$

and

$$q''_{2n}q''_{2n+1} = q'_{m_{2n-1}}q'_{m_{2n-1}+1} \cdots q'_{m_{2n+1}-1}, \qquad n = 0, 1, \ldots$$

(2) For $m = (0, 1, m_2, m_4, m_6, \ldots)$ and $m_o = (0, 1, 3, 5, \ldots)$ we have that $\varphi(m)(\Omega_{q_*}) = \varphi(m_o)(\Omega_{q''_*})$.
(3) For $m' = (0, m_1, m_3, m_5, \ldots)$ and $m_e = (0, 2, 4, 6, \ldots)$ we have that $\varphi(m')(\Omega_{q'_*}) = \varphi(m_e)(\Omega_{q''_*})$.

From the last properties, we conclude that

$$\varphi^{-1}(m') \circ \varphi(m_e) \circ \varphi^{-1}(m_o) \circ \varphi(m)$$

is a homeomorphism that is a conjugacy between the adding machine defined on Ω_{q_*} and the one in $\Omega_{q'_*}$. Let us remark that this, in fact, is a constructive proof of the conjugacy between multipermutative systems.

References

[1] V. Afraimovich, A. Maass, J. Urías, Symbolic dynamics for sticky sets in Hamiltonian systems, Nonlinearity 13 (2000) 1–21.

[2] V. Afraimovich, J. Schmeling, E. Ugalde, J. Urías, Spectra of dimensions for Poincaré recurrences, Discrete and Continuous Dynamical Systems 6 (2000) 901–914.

[3] V. Afraimovich, E. Ugalde, J. Urías, Poincaré recurrences for rhythmical dynamics, Far East J. Dynamical Systems 2 (2000) 53–67.

[4] V.S. Afraimovich, N.N. Verichev, M.I. Rabinovich, Stochastic synchronization of oscillations in dissipative systems, Izv. Vys. Uch. Zav. Radiofizika 29 (1986) 1050–1060, Radio Phys. and Quantum Electron 29 (1986) 747–751.

[5] V. Afraimovich, J.R. Chazottes, B. Saussol, Local dimensions for Poincaré recurrences, ERA of the AMS 6 (2000) 64–74, http://www.ams.org/era/home-2000.html.

[6] V. Afraimovich, J.R. Chazottes, E. Ugalde, Spectra of dimensions for Poincaré recurrences for special flows, Taiwanese J. Math. 6 (2002) 269–285.

[7] V. Afraimovich, J.R. Chazottes, B. Saussol, Pointwise dimension for Poincaré recurrences associated with maps and special flows, Discrete Contin. Dyn. Syst. 9 (2003) 263–280.

[8] V. Afraimovich, J. Urías, Some characteristics of complex behavior of orbits in dynamical systems, Commun. Nonlinear Sci. Numer. Simul. 8 (2003) 171–181.

[9] V. Afraimovich, Pesin's dimension for Poincaré recurrences, Chaos 7 (1997) 12–20.

[10] V. Afraimovich, Poincaré recurrences of coupled subsystems in synchronized regimes, Taiwanese J. of Math. 3 (1999) 139–161.

[11] V. Afraimovich, G.M. Zaslavsky, Sticky orbits of chaotic Hamiltonian dynamics, in: Lecture Notes in Physics, vol. 511, 1998, pp. 59–82.

[12] V. Afraimovich, G.M. Zaslavsky, Fractal and multifractal properties of exit times and Poincaré recurrences, Phys. Rev. E 55 (1997) 5418–5426.

[13] V.S. Afraimovich, W.-W. Lin, N.F. Rulkov, Fractal dimension for Poincaré recurrences as an indicator of synchronized chaotic regimes, Int. J. Bifurcations and Chaos 10 (2000) 2323–2337.

[14] V. I Arnold, V.S. Afraimovich, Yu.S. Il'yashenko, L.P. Shil'nikov, in: Encyclopedia of Mathematical Science, NS, Dynamical Systems, vol. V, Springer-Verlag, Berlin, Heidelberg, New York, 1994.

[15] L. Barreira, A non-additive thermodynamic formalism and applications to dimension theory of hyperbolic dynamical systems, Ergod. Th. & Dynam. Syst. 16 (1996) 871–928.

[16] L. Barreira, Ya. Pesin, J. Schmeling, On a general concept of multifractality: multifractal spectra for dimensions, entropies, and Lyapunov exponents. Multifractal rigidity, Chaos 7 (1997) 27–38.

[17] L. Barreira, Ya. Pesin, J. Schmeling, Dimension and product structure of hyperbolic measures, Annals Math. 149 (1999) 755–783.

[18] L. Barreira, J. Schmeling, Sets of "non-typical" points have full topological entropy and full Hausdorff dimension, Israel J. Math. 116 (2000) 29–70.

[19] L. Barreira, Hyperbolicity and recurrence in dynamical systems: a survey of recent results, Resenhas 5 (2002) 171–230.

[20] L. Barreira, B. Saussol, Product structure of Poincaré recurrences, Ergod. Th. & Dynam. Syst. 22 (2002) 33–61.

[21] S. Benkadda, G.M. Zaslavsky (Eds.), Chaos, Kinetics and Nonlinear Dynamics in Fluids and Plasmas, Lecture Notes in Physics, vol. 511, 1998.

[22] S. Benkadda, S. Kassibrakis, R. White, G.M. Zaslavsky, Self-similarity and transport in the standard map, Phys. Rev. E 55 (1997) 4909–4917.

[23] P. Billingsley, Ergodic Theory and Information, Wiley, New York, 1965.

[24] P. Billingslay, Hausdorff dimension in probability theory, Ill. J. Math. 4 (1960) 35–46.

[25] P. Billingslay, Hausdorff dimension in probability theory, II, Ill. J. Math. 5 (1961) 291–298.

[26] P. Billingsley, Probability and Measure, third ed., Wiley, 1995.

[27] T. Bohr, D. Rand, Entropy function for characteristic exponents, Physica 25D (1987) 387–398.

[28] M. Boshernitzan, Quantitative recurrence results. Rice Univ. preprint, 1991.

[29] R. Bowen, Topological entropy for non-compact sets, Trans. Amer. Math. Soc. 184 (1973) 125–136.

[30] R. Bowen, Equilibrium States and the Ergodic Theory of Anosov Diffeomorphisms, Lecture Notes in Mathematics, vol. 470, Springer-Verlag, Berlin, 1975.

[31] R. Bowen, Hausdorff dimension of quasi-circles, Publ. Math. IHES 50 (1979) 259–273.

[32] L. Breiman, The individual ergodic theorem of information theory, Annal. Math. Stat. 28 (1957) 809–811;
L. Breiman, Correction, Annal. Math. Stat. 31 (1960) 809–910.

[33] I.U. Bronshteyn, Synchronization of identical dynamical systems. Preprint, 1997.

[34] J.R. Brown, Ergodic Theory and Topological Dynamics, Academic Press, 1976.

[35] A.A. Brudno, Entropy and the complexity of the trajectories of a dynamical system, Trans. Moscow Math. Soc. 2 (1982) 127–151.

[36] H. Bruin, Dimensions of recurrence times and minimal subshifts, in: J.-M. Gambaudo, P. Hubert, P. Tisseur, S. Vaienti (Eds.), Dynamical Systems: From Crystal to Chaos, World Scientific, 2000, pp. 117–124.

[37] R. Cawley, R.D. Mauldin, Multifractal decompositions of Moran fractals, Adv. Math. 92 (1992) 196–236.

[38] J.R. Chazottes, F. Durand, Local rates of Poincaré recurrences for rotations and weak mixing, Discrete and Continuous Dyn. Sys. 12 (2005) 175–183.

[39] J.R. Chazottes, B. Saussol, On pointwise dimensions and spectra of measures, C. R. Acad. Sci. Paris Sér. I 333 (2001) 719–723.

[40] C. Carathéodory, Über das lineare Mass von Punktmengen – Eine Verallgemeinerung des Längenbegriffs, in: Gesammelte Mathematische Schriften, Band IV, 1914, pp. 249–277; English translation in: G.A. Edgar (Ed.), Classics on Fractals, Addison–Wesley, 1993, pp. 47–73 (The advanced book program).

[41] B.V. Chirikov, D.L. Shepelyanski, Correlation properties of dynamical chaos in Hamiltonian systems, Phys. D 13 (1984) 395–400;
B.V. Chirikov, Patterns in chaos, Chaos Solitons Fractals 1 (1991) 79–103.

[42] Z. Coelho, E. de Faria, Limit laws of entrance times for homeomorphisms of the circle, Israel J. Math. 93 (1996) 93–112.

[43] P. Collet, J.L. Lebowitz, A. Porzio, The dimension spectrum of some dynamical systems, in: Proceedings of the Symposium on Statistical Mechanics of Phase Transitions – Mathematical and Physical Aspects (Trebon, 1986), J. Statist. Phys. 47 (1987) 609–644.

[44] P. Collet, A. Galves, Asymptotic distribution of entrance times for expanding maps of an interval. Preprint, 1992.

[45] I.P. Cornfeld, S.V. Fomin, Ya. Sinai, Ergodic Theory, Springer-Verlag, Berlin, 1982.

[46] M.I. Cortez, F. Durand, B. Host, A. Maass, Continuous and measurable eigenfunctions of linearly recurrent dynamical Cantor systems, J. London Math. Soc. (2) 67 (2003) 790–804.

[47] P. Cvitanovic, M.H. Jensen, L.P. Kadanoff, I. Procaccia, Renormalization, unstable manifolds, and the fractal structure of mode locking, Phys. Rev. Lett. 55 (1985) 343–346;
T.C. Halsey, M.H. Jensen, L.P. Kadanoff, I. Procaccia, B.I. Schraiman, Fractal measures and their singularities: the characterization of strange sets, Phys. Rev. A 33 (1986) 1141–1151.

[48] W. de Melo, S. van Strien, One-Dimensional Dynamics, Springer-Verlag, Heidelberg, New York, 1993.

[49] M. Denker, C. Grillenberger, K. Sigmund, Ergodic Theory on Compact Spaces, Lecture Notes in Mathematics, vol. 527, Springer-Verlag, 1976.

[50] R.W. Easton, J.D. Meiss, et al., Exit times and transport for symplectic twist maps, Chaos 3 (1993) 153–165.

[51] R.S. Ellis, Entropy, Large Deviations and Statistical Mechanics, Springer-Verlag, 1985.

[52] B. Fernandez, E. Ugalde, J. Urías, Spectrum of dimensions for Poincaré recurrences of Markov maps, Discrete and Continuous Dyn. Sys. 8 (2002) 835–849.

[53] N. Fenichel, Persistence and smoothness of invariant manifolds of flows, Indiana Univ. Math. J. (1971) 173–226.

[54] U. Frish, G. Parisi, in: M. Ghill, R. Benzi, G. Parisi (Eds.), Turbulence and Predictability of Geophysical Flows and Climate Dynamics, North-Holland, Amsterdam, 1985.

[55] O. Frostman, Potential d'équilibre et capacité des ensembles avec quelques applications à la théorie des fonctions, Meddel. Lunds Univ. Math. Sem. 3 (1935) 1–118.

[56] H. Fujisaka, T. Yamada, Stability theory of synchronized motion in coupled-oscillator systems, Progress of Theoretical Physics 69 (1983) 32–47.

[57] H. Furstenberg, Disjointness in ergodic theory, minimal sets and a problem in Diophantine approximation, Mathematical Systems Theory 1 (1967) 1–49.

[58] H. Furstenberg, The structure of distal flows, Amer. J. Math. 85 (1963) 477–515.

[59] K. Gelfert, G.A. Leonov, V. Reitmann, Hausdorff dimension estimates by use of a tubular Carathéodory structure. Technische Universität Dresden, Preprint.

[60] J. Guckenheimer, P. Holmes, Nonliner Oscillations, Dynamical Systems and Bifurcations of Vector Fields, Springer-Verlag, New York, Heidelberg, Berlin, 1983.

[61] B.M. Gurevich, A.A. Tempelman, Hausdorff dimension and pressure in the DLR thermodynamic formalism, Amer. Math. Soc. Transl. 198 (2000) 91–107.

[62] J.K. Hale, Diffusive coupling, dissipation and synchronization, J. Dyn. Diff. Eq. 9 (1997) 1–52.

[63] P.R. Halmos, Ergodic theory, Math 373 A. Dept. of Math., Univ. of Chicago, Summer 1955.

[64] F. Hausdorff, Dimension und äußeres Maß, Mathematishe Annalen 79 (1918) 157–179; English translation in: G.A. Edgar (Ed.), Classics on Fractals, Addison–Wesely, 1993, pp. 75–100 (The advanced book program).

[65] R.H. Herman, I. Putnam, C.F. Skau, Ordered Bratteli diagrams, dimension groups and topological dynamics, Internat. J. Math. 3 (1992) 827–864.

[66] H.G.E. Hentschel, I. Procaccia, The infinite number of generalized dimensions of fractals and strange attractors, Physica D 8 (1983) 435–444;
P. Grassberger, I. Procaccia, Dimensions and entropies of strange attractors from a fluctuating dynamics approach, Physica D 13 (1984) 34–54.

[67] M. Hirata, Poisson law for Axiom A diffeomorphisms, Ergod. Th. & Dynam. Syst. 13 (1993) 533–556.

[68] M. Hirata, B. Saussol, S. Vaienti, Statistics of return times: a general framework and new applications, Commun. Math. Phys. 206 (1999) 33–55.

[69] V. Jarnik, Über die simultanen Diophantischen Approximation, Math. J. 33 (1931) 505–543.

[70] M. Kac, in: Lectures in Applied Mathematics, Interscience, London, 1957.

[71] M. Kac, On the notion of recurrence in discrete stochastic processes, Bull. Amer. Math. Soc. 53 (1947) 1002–1010.

[72] S. Kassibrakis, S. Benkadda, R.B. White, G.M. Zaslavsky, Nonuniversality of transport for the standard map, in: Lecture Notes in Physics, vol. 511, 1998, p. 403.

[73] A. Katok, B. Hasselblatt, Introduction to the Modern Theory of Dynamical Systems, Encyclopedia of Mathematics and its Applications, vol. 54, Cambridge University Press, London, New York, 1995.

[74] L. Kocarev, U. Parlitz, Generalized synchronization, predictability, and equivalence of unidirectionally coupled dynamical systems, Physical Review Letters 76 (1996) 1816–1819.

[75] A.N. Kolmogorov, S.V. Fomin, Elementos de la teoría de funciones y del análisis funcional, 2a edición, Editorial Mir, 1975.

[76] P. Kurka, A. Maass, Recurrence dimension in Toeplitz subshifts, in: J.M. Gambaudo, P. Hubert, P. Tisseur, S. Vaienti (Eds.), Dynamical Systems: From Crystal to Chaos, World Scientific, 2000, pp. 165–175.

[77] P. Kurka, Local return rates in substitutive subshifts, Acta Univ. Carolin. Math. Phys. 44 (2003) 29–42.

[78] P. Kurka, V. Penné, S. Vaienti, Dynamically defined recurrence dimension, Discrete & Cont. Dyn. Syst. 8 (2002) 137–146.

[79] M. Li, P. Vitányi, An Introduction to Kolmogorov Complexity and its Applications, Graduated Text in Computer Sciences, Springer-Verlag, New York, 1997.

[80] B. Mandelbrot, The Fractal Geometry of Nature, Freeman, San Francisco, 1982.

[81] L. Markus, K.R. Meyer, Periodic orbits and solenoids in generic Hamiltonian dynamical systems, Amer. J. Math. 102 (1980) 25–92.

[82] H. McCluskey, A. Manning, Hausdorff dimension for horseshoes, Ergod. Th. & Dynam. Syst. 3 (1983) 251–260.

[83] J.D. Meiss, Class renormalization: Islands around islands, Phys. Rev. A 34 (1986) 2375–2383.

[84] J.D. Meiss, Symplectic maps, variational principles and transport, Rev. Mod. Phys. 64 (1992) 795–848.

[85] J.D. Meiss, Average exit time for volume-preserving maps, Chaos 7 (1997) 139–147.

[86] J.H. Misguich, J.-D. Reuss, Y. Elskens, R. Balescu, Motion in a stochastic layer described by symbolic dynamics, Chaos 8 (1998) 248–256.

[87] P.A.P. Moran, Additive functions of intervals and Hausdorff measure, Proceedings of the Cambridge Philosophical Society 42 (1946) 15–23.

[88] A.D. Morozov, Quasi-Conservative Systems: Cycles, Resonances and Chaos, Nonlinear Science Series A, vol. 30, World Scientific, Singapore, 1998.

[89] V.V. Nemyckii, V.V. Stepanov, Quantitative Theory of Differential Equations, Princeton University Press, Princeton, NJ, 1960.

[90] Z. Nitecki, Differentiable Dynamics, MIT Press, Cambridge, MA, 1971.

[91] D. Ornstein, B. Weiss, The Shannon–McMillan–Breiman theorem for amenable groups, Israel J. Math. 44 (1983) 53–60.

[92] G. Paladin, A. Vulpani, Anomalous scaling laws in multifractal objects, Phys. Rep. 156 (1987) 147–225.

[93] L.M. Pecora, T.L. Carrol, G.A. Johnson, D.J. Mar, J.F. Heagy, Fundamentals of synchronization in chaotic systems, concepts and applications, Chaos 7 (1997) 520–543.

[94] L.M. Pecora, T.L. Carroll, Synchronization in chaotic systems, Physical Review Letters 64 (1990) 821–824.

[95] V. Penné, B. Saussol, S. Vaienti, Dimensions for recurrence times: topological and dynamical properties, Discrete and Continuous Dynamical Systems 5 (1999) 783–798.

[96] V. Penne, B. Saussol, S. Vaienti, Fractal and statistical characteristics of recurrence times. Talk at the conference "Disorder and Chaos", Rome (Sept. 1997). CPT (Luminy) preprint, 1997.

[97] Ya. Pesin, Dimension Theory in Dynamical Systems: Contemporary Views and Applications, Chicago Lectures in Mathematics, The University of Chicago Press, 1997.

[98] Ya. Pesin, Dimension type characteristics for invariant sets of dynamical systems, Russ. Math. Surveys 43 (1988) 111–151.

[99] Ya. Pesin, Generalized spectrum for dimensions: the approach based on Carathéodory construction, in: Th.M. Rassias (Ed.), Constantin Carathéodory: An International Tribute, World Scientific, Singapore, 1991, pp. 1108–1119.

[100] Ya. Pesin, H. Weiss, A multifractal analysis of equilibrium measures for conformal expanding maps and Markov Moran geometric constructions, J. Stat. Phys. 86 (1–2) (1997) 233–275.

[101] Ya. Pesin, H. Weiss, The multifractal analysis of Gibbs measures: motivation, mathematical foundation, and examples, Chaos 7 (1997) 89–106.

[102] A.S. Pikovsky, M.G. Rosenblum, G.V. Osipov, J. Kurths, Phase synchronization of chaotic oscillators by external driving, Physica D 104 (1997) 219–238.

[103] B. Pitskel, Poisson limit law for Markov chains, Ergod. Th. & Dynam. Syst. 11 (1991) 501–513.
[104] M. Pollicott, M. Yuri, Dynamical Systems and Ergodic Theory, Cambridge University Press, Cambridge, 1998.
[105] M. Rabinovich, Private communication.
[106] A. Rényi, Probability Theory, North-Holland, Amsterdam, 1970.
[107] D. Ruelle, Thermodynamic formalism for maps satisfying positive expansiveness and specification, Nonlinearity 5 (6) (1992) 1223–1236.
[108] D. Ruelle, Repellers for real analytic maps, Ergod. Th. & Dynam. Syst. 2 (1982) 99–107.
[109] D. Ruelle, Thermodynamic Formalism, Encyclopedia of Mathematics and its Applications, vol. 5, Addison–Wesley, 1978.
[110] N.F. Rulkov, M.M. Sushchik, L.S. Tsimring, H.D.I. Abarbanel, Generalized synchronization of chaos in directionally coupled chaotic systems, Phys. Rev. E 51 (1996) 980.
[111] J. Schmeling, H. Weiss, An overview of the dimension theory of dynamical systems, in: Proc. Sympos. Pure Math., vol. 69, Amer. Math. Soc., Providence, RI, 2001, pp. 429–488.
[112] B. Saussol, On fluctuations and the exponential statistics of return times, Nonlinearity 14 (1) (2001) 179–191.
[113] B. Saussol, S. Troubetzkoy, S. Vaienti, Recurrence, dimensions, and Lyapunov exponents, J. Stat. Phys. 106 (3/4) (2002) 623–634.
[114] B. Saussol, S. Troubetzkoy, S. Vaienti, Recurrence and Lyapunov Exponents, Mosc. Math. J. 3 (1) (2003) 189–203.
[115] A. Tempelman, Dimension of random fractals in metric spaces, Theory of Probabilities and Applications, No. 2 (1999).
[116] E.B. Vul, Ya.G. Sinai, K.M. Khanin, Feigenbaum universality and thermodynamic formalism, Russian Mat. Surveys 39 (3) (1984) 1–30.
[117] X.-J. Wang, Statistical physics of temporal intermittency, Phys. Rev. A 40 (1989) 6647–6661.
[118] P. Walters, A variational principle for the pressure of continuous transformations, Amer. J. Math. 97 (1976) 937–971.
[119] P. Walters, Invariant measures and equilibrium states for some mappings which expand distances, Trans. Amer. Math. Soc. 236 (1978) 121–153.
[120] P. Walters, An Introduction to Ergodic Theory, Springer Graduate Texts in Math, vol. 79, Springer, New York, 1982.
[121] H.S. White, Algorithmic complexity of points in dynamical systems, Ergod. Th. & Dynam. Sys. 13 (1993) 807–830.
[122] C.W. Wu, L.O. Chua, A unified framework for synchronization and control of dynamical systems, Int. J. of Bifurcation and Chaos 2 (1994) 979–988.
[123] L.-S. Young, Recurrence times and rates of mixing, Israel J. of Math. 110 (1999) 153–188.

[124] L.-S. Young, Dimension, entropy and Lyapunov exponents, Ergod. Th. & Dynam. Syst. 2 (1982) 109–124.

[125] V.F. Sirvent, Spectra of recurrence dimension for adic systems, Stoch. Dyn. 2 (2002) 599–607;
V. Sirvent, Spectra of recurrence dimensions for dynamically defined subsets on Rauzy fractals, Fractals 14 (1) (2006) 1–9.

[126] G.M. Zaslavsky, M. Edelman, B. Niyazov, Self-similarity, renormalization, and phase space nonuniformity of Hamiltonian chaotic dynamics, Chaos 7 (1997) 159–181.

[127] G.M. Zaslavsky, Physics of Chaos in Hamiltonian Systems, Imperial College Press, London, Singapore, 1998.

[128] G.M. Zaslavsky, Hamiltonian Chaos and Fractional Dynamics, Oxford University Press, Oxford, 2005.

Subject Index